D0843399

Experimental Design
in Biotechnology

STATISTICS: Textbooks and Monographs

A Series Edited by

D. B. Owen, Coordinating Editor
Department of Statistics
Southern Methodist University
Dallas, Texas

R. G. Cornell, Associate Editor
for Biostatistics
University of Michigan

W. J. Kennedy, Associate Editor
for Statistical Computing
Iowa State University

A. M. Kshirsagar, Associate Editor
for Multivariate Analysis and
Experimental Design
University of Michigan

E. G. Schilling, Associate Editor
for Statistical Quality Control
Rochester Institute of Technology

ADDITIONAL VOLUMES IN PREPARATION

Experimental Design in Biotechnology

Perry D. Haaland

Becton Dickinson and Company
Research Center
Research Triangle Park, North Carolina

MARCEL DEKKER

New York

ISBN 0-8247-7881-2

This book is printed on acid-free paper.

MARCEL DEKKER
270 Madison Avenue, New York, New York 10016

Current printing (last digit):
10 9 8 7

PRINTED IN THE UNITED STATES OF AMERICA

Dedicated to Pamela and Wren

PREFACE

This book is intended to appeal to researchers who, because of limited time and resources, must use small, efficient experiments to solve complex biological problems. It is written to provide the first time user of statistics with an understanding of how and why statistical experimental design and analysis can be an effective problem solving tool. The presentation is by example, and the approach and methods are graphical rather than numerical. All of the statistical methods can be performed with readily available personal computer statistics packages.

Statistical experimental design and analysis can increase research productivity because it helps the researcher to

- ask the right questions,
- collect data which can answer the questions, and
- analyze the data to reveal the answers.

The best way to learn these skills is in the context of real examples. Thus, biotechnology examples such as the development of stability enhancing buffers for immunoassays, optimization of monoclonal antibody production and the coating of surfaces to enhance biological activity are used to present and illustrate the statistical methods.

The experimental designs presented in this book are useful for small screening and response surface experiments. Screening experiments are widely used in industrial research to identify which few (out of many) process variables have an important effect on process performance and to determine how changing the settings of these important variables can improve performance. Response surface experiments are used to optimize biological processes and technologies. Because resources for such experiments are limited, the focus of this work is on small experiments. A tabulation of useful designs for these experiments is included in the Design Digest at the end of this book.

This text has grown out of my consulting practice and a series of short courses offered to researchers at the Becton Dickinson Research Center. These experiences taught me that statistical methods are most useful as part of an overall problem solving strategy. Therefore, I have attempted to preserve the balance between the scientific, problem solving aspects of the examples and the statistical analysis. To further develop this relationship, two case histories of successful statistical problem solving are included in the text.

This book is intended to be of practical use. The experimental designs were selected because they are practical and easy to use. The important properties of the designs and how to select the best design are clearly explained. To complement the practical use of experimental design, the statistical analysis is centered on readily understood, widely useful graphical methods which can serve for both analysis and presentation. Finally, the real problem solving examples should help the reader identify likely applications in his or her own work.

The main emphasis of this work is on problem solving, and statistical theory is presented only as necessary to clarify important points. Thus, this book is not intended as a replacement for texts on analysis of variance, regression analysis or more advanced treatments of experimental design. Readers who are interested in these topics will find resources in the references and bibliography sections at the end of each chapter. The well-known work of Box, Hunter and Hunter, *Statistics for Experimenters* (1978), is highly recommended for further reading and study.

Acknowledgments

A special debt is gratefully acknowledged to the many researchers at the Becton Dickinson Research Center (BDRC) whose work provided the experience from which this book grew. Robert Adrion, Frank Blinkhorn, Beverly Fleming, Randy Hoke, Roger Liddle, Al Manson, Jim Mapes, Candace Weck, Gary Seibert, Nancy Shields, and Doreen Yen provided examples used in the text. Support for the BDRC Statistics Group and a research environment encouraging the productive use of statistics were provided by BDRC Director Roger Wilsnack.

Roger Liddle prepared the Design Digest, provided examples from his consulting and was a continual source of advice and help throughout the preparation of this book. Robert Adrion, formerly my departmental manager, served as my advocate and supporter in applying statistical problem solving methods at BDRC. Conrad Fung's comments on a draft of this manuscript led to many clarifications and improvements in the presentation. Finally, special thanks to my wife and daughter, who provided the balance and love which made this work possible.

This book was produced on a Sun™ workstation using Frame Maker™ software. Statistical graphics were produced using the SPlus™ statistical analysis system. (Sun is a trademark of Sun Microsystems, Inc. Frame Maker is a trademark of Frame Technology Corporation. SPlus is a trademark of Statistical Sciences, Inc.)

Perry D. Haaland

CONTENTS

Experimental Design
in Biotechnology

Chapter 1

STATISTICAL PROBLEM SOLVING

For most bioprocess technologies there are no theoretical models which can be used to explain process performance. Consequently, successful research is characterized by effective empirical problem solving. Typically, the problem solving process is governed by limitations on time and resources. Therefore, research productivity is a critical concern. Statistical problem solving provides a set of powerful tools which can be used to maximize the efficiency and productivity of empirical problem solving.

1.1   Collecting information-rich data

Research in biotechnology generates great quantities of data. The recent spread of computers in the research environment has greatly increased our ability to collect and manage this data. However, although computers increase the amount of data we can create and manage, they do not necessarily increase the information in the data. Since there are limited time and resources available to generate and understand this data, it is important that the data be information-rich. Statistical experimental design is one way to increase the amount of information-rich data we collect.

1

Purpose of collecting data

Data is collected to solve empirical problems. Data serve as a basis both for understanding and for action. Some typical reasons for collecting data for empirical problem solving are as follows:

- determine which few out of many variables significantly affect process performance
- determine how the settings of the variables should be changed in order to improve the process performance
- determine the optimal process performance level and specify what actions must be taken to achieve this level.

These reasons for collecting data have long been a part of statistical methodology. Methods for their use were described, for example, in the books *Design and Analysis of Industrial Experiments* by Owen L. Davies (1956), *Applications of Statistics to Industrial Experimentation* by Cuthbert Daniel (1976), *Statistics for Experimenters* by George E. P. Box, William G. Hunter and J. Stuart Hunter (1978), and *Practical Experimental Design* by William Diamond (1981). In this book, we use the power of real examples to show how these statistical problem solving methods can be applied to the collection and analysis of data from biotechnology experiments.

How should data be collected?

Every experiment has a design. Some designs are better than others. Since data must be collected anyway, the use of statistically designed experiments for data collection adds only incrementally to its cost. However, well designed experiments significantly increase the information content of the data.

Since data serve as the basis for understanding and action, it is essential to have correct data to make correct decisions. In *Guide to Quality Control*, K. Ishikawa (1976) outlined the following considerations in collecting data:

- "Will the data determine the facts?"
- "Are the data collected, analyzed and compared in such a way as to reveal the facts?"

The first consideration relates to experimental design while the second deals with statistical analysis. However, data which are collected properly are usually simple to analyze and understand.

The first step in the proper collection of data is to clarify the objectives of the experiment. Next the experimenter must determine what data to collect, how to measure it, and how the data relate to process performance and experimental objectives. The experimenter must ensure that the data collected are representative of the process so that the data will lead to correct conclusions.

Finally, an experimental design must be chosen which will reveal the facts as they relate to the experimental objectives.

What is an experimental design?

An experimental design is a collection of predetermined settings of the process variables. Each process variable is called an experimental factor. Each combination of settings for the process variables is called a run. A response variable is a measure of process performance. Each value of the response variable is called an observation. For example, for a process which depends on pH and incubation time, a possible experimental design is as follows:

- run 1: pH=6.7, time=20 minutes;
- run 2: pH=6.7, time=30 minutes;
- run 3: pH=7.2, time=20 minutes;
- run 4: pH=7.2, time=30 minutes.

At each of these runs, process performance would be observed. The resulting data set would provide information about how pH and time affect process performance and what can be done to improve process performance.

Progress on the learning curve

Empirical problem solving is difficult because the problems are complex and progress along the learning curve is often slow and difficult. The use of statistical experimental design speeds progress along the learning curve because these experiments are very effective in answering questions. (See Figure 1.1) This sounds simple, but anyone who has carried out an experiment, examined the results, and then was unable to provide any clear answers can appreciate the fundamental importance of this idea.

1.2   Separating signals from the noise

Data represent a combination of signals and noise. A signal is an indication of the effect that a process variable has on process performance. In a properly designed experiment, any factor which has an important effect on process performance will generate a signal which is large in comparison to the noise. Unimportant factors will have signals which fall below the noise level. The noise is all of the other variation in the data. Statistical methods can be used to separate the signals from the noise.

Noise

If we repeat a biological experiment, we don't expect to get exactly the same results as in the first experiment. This is true even if we are very careful to run

Knowledge

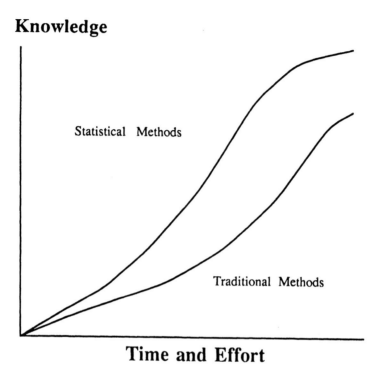

Statistical   Methods

Traditional   Methods

Time and Effort

Figure 1.1 A learning curve measures how much time and effort it takes to acquire knowledge about the problem being studied. Statistical methods provide an advantage over traditional problem solving methods such as "one-factor-at a-time" or haphazard experimentation because statistical methods focus on small, well designed experiments which answer questions about the process being studied. This results in a steeper learning curve and faster progress toward the problem solution.

each experiment under controlled conditions. The differences between experiments contribute to the uncertainty about the true results. Statistical methods allow us to reason in the presence of this uncertainty.

Uncertainty stems from variability in measurements of process performance. Variability can be thought of in terms of two components; namely, experimental error and measurement error. Measurement error is the variability observed when remeasuring the same experimental outcome. Experimental error is the difference in outcomes when an experiment is independently repeated under controlled conditions. Experimental error is usually much larger than measurement error. When we refer to noise, we are usually referring to the sum of both experimental and measurement error.

Clear signal designs

Experimenters can often identify many factors which may affect a process, but they usually have limited resources to carry out their experiments. Therefore, the use of small experiments is especially attractive. However, special care must be taken in using a small experiment under these conditions so that meaningful results will be achieved.

For example, in formulating a new assay buffer, the experimenter needs to be confident that, say, the effect of pH can be clearly separated from the effect of ionic strength. If, for the purposes of economy, the process is only observed with pH and ionic strength both at low levels or both at high levels, then the effects of these two factors cannot be separated. However, if all four combinations of high and low levels were observed, then the effects of the two factors would be clearly separated.

In order to conduct efficient experiments, we want to study as many experimental factors as we can with the fewest number of observations. As the number of factors increases, it becomes more difficult to choose combinations of their levels which clearly separate their effects yet at the same time minimize the size of the experiment.

The Design Digest, at the end of this book, is a collection of efficient experimental designs which allow many factors to be investigated in small experiments. These designs can be characterized as "clear signal" designs (J. Stuart Hunter, 1987). That is, when used properly, they

- separate the signals from the noise and
- clearly distinguish signals from each other.

These designs form the basis for efficient, effective statistical problem solving.

The Pareto Principle

A statistical analysis provides estimates of how strongly each experimental factor affects process performance. These estimates reveal which factors are most important and how changing their settings affects process performance. An interesting way to compare the relative importance of the estimated effects of the factors is by means of a Pareto chart (Ishikawa, 1976).

The Pareto chart in Figure 1.2 graphically displays the magnitudes of the effects from a data set which is analyzed in Chapter 4. The effects are sorted from largest to smallest. The Pareto chart clearly shows that the first effect is by far the most important one. The second effect also seems to be above the

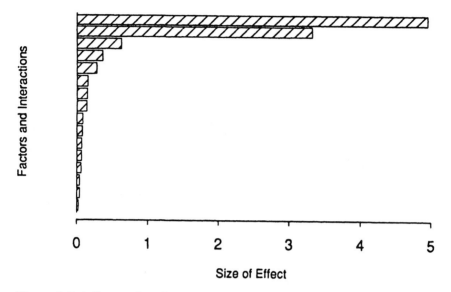

Figure 1.2 A Pareto chart for the results of an experimental design graphically depicts the relative magnitudes of the effects of each of the experimental factors. The y-axis shows the experimental factors (and possibly their interactions). The x-axis shows the absolute magnitude of the effect of each factor as determined by the statistical analysis. The Pareto Principle suggests that most of the improvement the process performance can be obtained by changes in only a few of the experimental factors.

noise level but is much smaller. In Chapter 4, statistical methods are presented for determining which, if any, of the remaining effects are important.

To improve process performance, it is obvious that we should begin by adjusting the factor which had the biggest effect. The Pareto chart is an excellent tool for identifying this factor. It also helps us to have a clear understanding of the relative effects of the rest of the experimental factors, regardless of their statistical significance.

J. M. Juran, in his book *Managerial Breakthrough* (1964), described the Pareto Principle; namely that process variables can be classified as belonging to either "the vital few" or "the trivial many". Improvements in process performance come from paying attention to the few really important factors which affect the process. Another way to think of this is according to the "80/20" rule which says that 80% of the improvement in process performance will come from changes in 20% of the process variables.

In order to find these "vital few" process variables, a statistical experimental design is used which can provide a clear estimate of the effect of each factor on process performance. Then data, which allow us to estimate the effects, is collected according to the design. A Pareto chart will usually point out clearly the "vital few" effects with which we should concern ourselves. The statistical analysis tells us which factors rise above the noise level. (However, since the magnitudes of the calculated effects depend on the choice of experimental ranges, if we incorrectly choose the range for an important effect, it may not show up as an important variable on the Pareto chart or in the statistical analysis.)

The Pareto Principle is analogous to the statistical principle of *factor sparsity* (Box and Meyer, 1986). Factor sparsity assumes that "in relation to the noise only a small proportion of the factors have effects that are large"; that is, a few factors are "active" and the rest are "inert". For our examples, factor sparsity generally implies *effect sparsity*. Effect sparsity assumes that, out of the possibly many calculated effects due to the factors and their interactions, only a few will be important. Thus, the use of a Pareto chart to identify the few large calculated effects depends on effect sparsity. The experimental designs discussed in this book are especially good for finding the few most important factors and their associated effects.

1.3   Biotechnology applications

Although statistical experimental design has been widely used in many areas of science and industry (see, for example, Box, Hunter and Hunter, 1978; Daniel, 1976; Diamond, 1981; or Taguchi, 1986), it has not yet been widely adopted in the biological sciences. We believe this is partly because of the lack of a clear explanation of the methods using biotechnology examples. This book addresses this need by showing examples of how statistical problem solving is used effectively in real biotechnology applications.

The applications we discuss require effective problem solving methods because they involve

- many factors
- no theoretical model
- data which may include high levels of noise and
- possible interactions among variables.

This description fits many problems in the development of bioprocess technologies. Indeed, the reader can easily see the general applicability of these methods to scientific research.

Examples

The researchers with whom we work successfully use statistical problem solving in many projects. In fact, their experiences form the basis of this book. Among the real-life examples we use to illustrate the use of statistical methods are the following problems:

- formulate a storage buffer for an enzyme-immunoassay which enhances its stability
- configure an enzyme-immunoassay so as to maximize its signal-to-noise ratio
- identify treatment factors which affect the activity of a bioactive compound coating a polyurethane surface
- optimize production of monoclonal antibodies harvested from ascites producing mice
- identify important factors affecting the yield of monoclonal antibodies produced by a cell culture system.

In all of these examples, statistical problem solving methods were successfully used to find practical solutions. The statistical methods were learned and used by the researchers themselves.

The use of statistical experimental design increases the efficiency and productivity of the research experiments. The statistical analysis provides a powerful framework within which to ask and answer questions about possible solutions. Finally, the systematic problem solving strategy allows the scientists to effectively integrate all of these tools into their research.

The statistical approach to solving empirical problems provides a common language which scientists, engineers and managers can use to communicate in planning, carrying out, and evaluating experimental programs. Even more importantly, it provides a means by which problem solvers can speed up their progress along the learning curve. Real examples make these powerful tools more readily accessible to biotechnology researchers.

1.4   Comparison of strategies for problem solving

Successful problem solving consists of

- asking good questions,
- collecting data which can answer the questions, and
- analyzing the data to reveal the answers.

Problem solving can be made more productive by using small, efficient experimental designs when collecting data and by using powerful statistical graphics and analysis to understand the results (Figure 1.3).

One-at-a-time and matrix methods

Two methods which are sometimes used as alternatives to statistical experimental design and analysis are "one-at-a-time" and "matrix" experimentation. Each of these methods can be used for problem solving, but neither of them is economical or efficient.

The "one-variable-at-a-time" approach is to fix all of the variables except one and then study the behavior of the system at several levels of that variable. For each variable the best value is found, and then the process is repeated for the remaining variables until all variables have been considered. Figure 1.4a illustrates this approach. This method may be effective in some situations, but it is very inefficient. It just takes too many experiments to come up

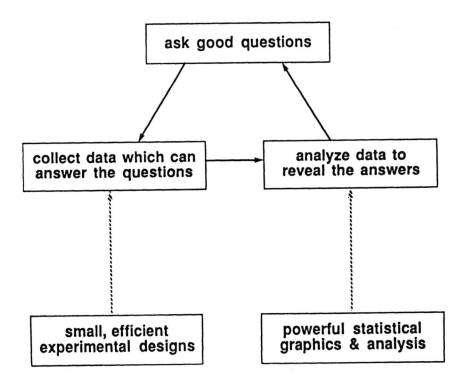

Figure 1.3 Problem solving consists of a cycle of asking questions, collecting data which can answer the questions, and analyzing the data to reveal the answers. The effectiveness of this process can be greatly increased by using statistical experimental design to collect better data and statistical analysis to more clearly reveal the answers.

with an answer. If there are interactions among the variables the "one-at-a-time" method may miss the solution because it doesn't thoroughly explore the space of possible solutions.

A second traditional approach to empirical problem solving is to lay out a matrix of all interesting combinations of the variables being investigated. Then all of the combinations in the matrix are investigated until the solution is found. This method is illustrated in Figure 1.4b. The matrix method has the advantage of thoroughly exploring the experimental space, but it requires an unnecessarily large number of measurements. Even with only four or five variables the matrix is too large to be explored in a realistic amount of time.

Statistical design

The statistical problem solving approach uses a series of small, carefully designed experiments. Each experiment carefully explores the experimental space while studying many variables using a small number of observations. This method is illustrated in Figure 1.4c. For each small experiment, well defined questions are asked, and simple statistical methods are used to provide answers. A clear strategy is used to ensure efficient progress toward a solution.

1.5   Iterative problem solving strategy

We sometimes call the statistical design approach "strategic experimentation" because it couples statistical experimental design and analysis with a well defined problem solving strategy. An important characteristic of this approach is that it is iterative. By this we mean that we use a series of small experiments to solve difficult problems. We also call this the

"Stop, Look, and Listen"

approach to experimentation. That is, we do a small experiment, learn from the results, and then plan the next experiment. (See Figure 1.5.) This iterative procedure refines and improves our questions to take advantage of what has been learned so far.

When planning an experiment, we have different objectives depending on what is known so far about the solution to the problem. If we are just beginning, there may be a long list of possible variables or factors which may affect the process. We may only have a preliminary idea of the reasonable ranges for the factor values. On the other hand, if we are close to finding a solution there may only be a few factors which are known to be important. Finally, at the conclusion of an investigation, the results are confirmed in a follow-up experi-

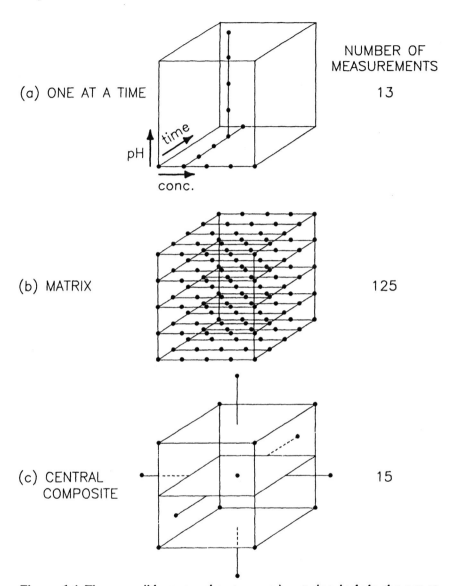

(a) ONE AT A TIME   13

(b) MATRIX   125

(c) CENTRAL
    COMPOSITE   15

Figure 1.4 Three possible approaches to experimentation include the one-at-a-time approach (a), the matrix method (b) and the statistical design approach (c). The one-at-a-time method doesn't require many measurement, but it doesn't explore the experimental space very well and may miss the solution. The matrix approach is effective but inefficient because it requires too many measurements. The statistical design method is efficient and effective because it provides good coverage of the experimental space with as few measurements as possible (Adrion, *et al*, 1984).

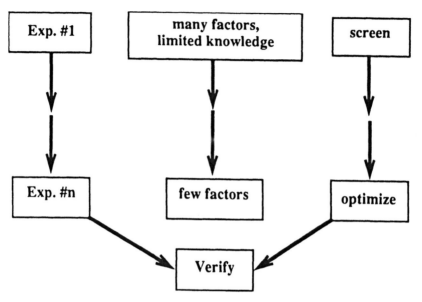

Figure 1.5 An iterative approach to solving problems involves thinking careful-ly about the problem, designing an experiment, collecting the data, and analyz-ing the results. This process may be repeated as necessary until the problem is solved.

ment. These three stages of experimentation are called, respectively, screen-ing, optimization, and verification.

A useful analogy to the idea of iterative problem solving is climbing a mountain. For example, we don't want to expend most of our resources just getting to the base camp. At timely intervals, we want to evaluate whether or not the current direction is leading us toward the summit. Different climbing strategies are required for the final ascent to the summit than for earlier stag-es of the climb. In this context, screening experiments help us identify plausi-ble directions to climb in order to reach the summit, optimization experiments are the ascent to the summit, and verification experiments prove we were there. Thus, the iterative nature of strategic experimentation allows us to adapt our methods to the changing nature of our objectives.

Using small experiments

If we are going to do several experiments, then each experiment has to be fair-ly small. In fact, given fixed resources, it is usually better to do several small experiments rather than one large one because the information from one experiment can improve the design of the next experiment. A valuable rule of thumb is to not use more than 25% of our resources for the first experiment (Box, Hunter and Hunter, 1978).

It is usually a good idea for a first experiment to be as small as possible. For one reason, initial guesses of the best settings of the factors may not be very close to their best settings. Some of the experimental conditions may not be practical or simply don't work. After looking at the data, we may decide that an important factor was left out of the design. On the other hand, the initial guesses for the important factors and their values may have been good enough to suggest that we should go directly to an optimization experiment.

Screening - optimization - verification

One purpose of doing a series of small experiments is that, as the investigation progresses, we can change the kind of experiment we do in order to reflect

	Objective	**Description**
Screening	identify important factors	many factors imprecise knowledge
Optimization	build predictive model	few factors in region of optimum
Verification	confirm results	at predicted best settings

Figure 1.6 Screening and optimization experiments have different objectives. Screening experiments are used early in an investigation to narrow the focus of the problem. Optimization experiments are used at the end of an investigation to build a predictive model which can be used to provide specific information about the solution. A verification experiment concludes the investigation.

changing objectives. These different experiments are classified as screening, optimization and verification experiments. Figure 1.6 provides a brief description of the conditions under which these three types of experiments are used.

Screening experiments are small experiments which include many variables. They play an important role in the early stages of an investigation. Because they are small and include many factors, they don't have as much information per factor. Their objective is problem reduction; that is, to focus on the important variables and to find out more about their best settings.

The purpose of an optimization experiment is to build a mathematical model which can be used to predict the behavior of the process being investigated. This requires a lot of information about each factor so that optimization experiments usually only include a few factors but are fairly large compared to

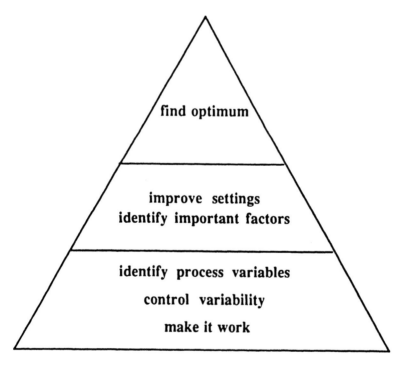

Figure 1.7 At the foundation of statistical problem solving is knowledge of the scientific aspects of the problem. Next, important factors are identified through screening factors. Optimization experiments are used to find the best process performance. The culmination is an experiment which verifies that the optimum process performance has been achieved.

screening experiments. Their objective is to produce specific optimal values for the experimental factors.

The simplest type of verification experiment shows that the predicted optimal process performance can be reproduced in a second experiment. This may involve production runs or further laboratory experiments. A second, larger type of verification experiment may be designed to verify that, over a given range, a set important factors have the predicted effect on the process being studied.

Building a solution

One way to think of empirical problem solving is that we are trying to build a solution to the problem (Figure 1.7); that is, we lay a solid foundation and then systematically add layers of further information. We start by thinking about which factors may effect process performance, which factors contribute to process variability, and how we can utilize this information to get the process working at a basic level. This information provides the basis for moving to the screening stage in which we try to identify which of the (possibly many) factors are important and to discover how we can change the factor settings to improve process performance. The culmination of the problem solving process is finding the optimal settings of the most important factors.

1.6   Conclusion

Problem solving is a learning experience. This learning occurs as we reduce the uncertainty about the problem. Beginning with screening experiments and proceeding to verification experiments, we reduce uncertainty by determining which out of many factors have an important effect on the process and what are the best settings of the important factors. An illustration of how our knowledge about the problem increases, and hence how our uncertainty decreases, is given in Figure 1.8.

The successful problem solving strategy which we use has the following characteristics:

- Data is collected with clear objectives in mind in order to insure that the data will determine the important facts about the process being studied.
- Information-rich data is insured by the use of statistical experimental designs for data collection.
- Answers to questions are provided by appropriate statistical analysis and graphical displays.
- A series of small experiments is used in an iterative approach to solving complex biological problems.

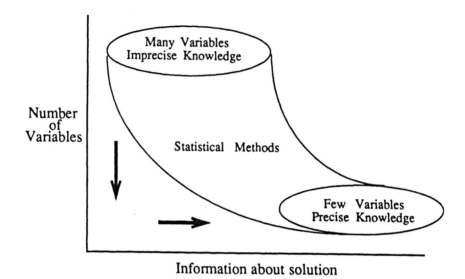

Information about solution

Figure 1.8 Two variables which affect uncertainty about the solution to an empirical problem are which factors are important (y-axis) and how much is known about their best values (x-axis). When we first approach an empirical problem we are in the upper left hand corner of the x-y graph. The solution to the problem lies in the lower right hand corner. Statistical problem solving is an efficient and effective method for taking us there.

Although these problem solving methods have a long and successful history in many areas of science, resistance to change is universal. We have heard people express their resistance to changing their problem solving approach in some of the following sayings:

- "We'll worry about the statistics after we've run the experiment."
- "Let's just vary one thing at a time so that we don't get confused."
- "I'll include that factor in the next experiment."
- "There aren't any interactions."
- "It's too early to use statistical methods."
- "A statistical experiment would be too large."
- "My data are too variable to use statistics."

However, these reasons are precisely why statistical problem solving tools should be used.

1.7 References and Bibliography

Adrion, R. F., G. R. Siebert, C. J. Weck, D. Yen, and A. R. Manson (1984). Optimization of in vivo monoclonal antibody production using computer-assisted experimental design. *Proceedings of the First Carolina Biomedical Engineering Conference*, North Carolina Biotechnology Center, P. O. Box 12235, Research Triangle Park, NC 27709, 125-144.

Box, G. E. P. and N. R. Draper (1969). *Evolutionary Operation*. New York: Wiley.

Box, G. E. P. and N. R. Draper (1987). *Empirical Model-Building and Response Surfaces*. New York: Wiley.

Box, G. E. P., W. G. Hunter, and J. S. Hunter (1978). *Statistics for Experimenters: An Introduction to Design, Data Analysis, and Model Building*. New York: Wiley.

Box, G. E. P. and R. D. Meyer (1986). An analysis for unreplicated fractional factorials. *Technometrics*, 28, No. 1, 11-18.

Cornell, J. A. (1980). *Experiments with Mixtures: Designs, Models and the Analysis of Mixture Data*. New York: Wiley.

Daniel, C. (1976). *Applications of Statistics to Industrial Experimentation*. New York: Wiley.

Davies, O. L. (1956). *Design and Analysis of Industrial Experiments*, 2nd ed. New York: Wiley.

Deming, S. N. and S. L. Morgan (1987). *Experimental Design: A Chemometric Approach*. New York: Elsevier.

Diamond, W. J. (1981). *Practical Experimental Designs*. Belmont, CA: Lifetime Learning Publications.

Hicks, C. R. (1983). *Fundamental Concepts in the Design of Experiments*, 3rd ed. New York: Holt, Rinehart and Winston.

Hunter, J. S. (1987). Experimental design: A winning strategy for industry. Seminar series sponsored by Bolt, Beranek and Newman.

Ishikawa, K. (1976). *Guide to Quality Control*. New York: Unipub.

Juran, J. M. (1964). *Managerial Breakthrough*. New York: McGraw-Hill.

Khuri, A. I., and J. A. Cornell (1987). *Response Surfaces: Designs and Analyses*. New York: Marcel Dekker, Inc. and Milwaukee: ASQC Quality Press.

Montgomery, D. C. (1976). *Design and Analysis of Experiments.* New York: Wiley.

Myers, R. H. (1976). *Response Surface Methodology.* Ann Arbor, Mich.: Edwards Brothers (distributors).

Taguchi, G. (1986). *Introduction to Quality Engineering: Designing Quality into Products and Processes.* Tokyo: Asian Productivity Organization.

Chapter 2

OPTIMIZATION OF *IN VIVO* PRODUCTION OF MONOCLONAL ANTIBODIES

In this chapter, we present a real example which illustrates the statistical problem solving methodology. The problem involves the optimization of *in vivo* production of monoclonal antibodies (Adrion, Siebert, Weck, Yen and Manson, 1984). The researchers who carried out this investigation used statistical methods for their efficiency, power and productivity.

2.1 Problem description

In this production process, large quantities of monoclonal antibodies may be prepared by growing antibody secreting cells in mice and then collecting an antibody-rich fluid called ascites. The antibody secreting cells are all identical twins of a single cell which was originally selected using monoclonal antibody technology. This original cell and all of its twins (clones) produce identical antibody molecules so the antibodies are called monoclonal. Once the cell line is established, animals (e.g., mice) may be used as hosts to support further growth of the cell line. The antibodies can then be harvested by collecting the fluid (ascites) which is continuously secreted by the cells.

Problem solving approach

In early stages of monoclonal antibody development, it is common for process variables to be set based on personal experience and preference. However, in order to insure the commercial viability of such production, a more rigorous approach to optimizing the production process is required.

Many variables may potentially affect the efficiency of such a production process, and there are distinct possibilities that some of the variables investigated may interact to produce unexpected results. Since there is no theoretical model for this complex biological process, the problem must be approached by experimentation. The complexity of the physiological processes involved implies that experimental results are subject to inherent variability.

The experimentation required to solve such a problem is expensive and time consuming. Each experiment involves many mice which must be housed and cared for during the several weeks required for the mice to begin producing antibodies. Thus, it is important to minimize the number of experimental conditions which must be investigated. The lengthy time between starting an experiment and finding out the results means that we want to conduct as few experiments as possible. Based on these constraints, we adopt the following approach:

- use a screening experiment to identify important factors,
- use an optimization experiment to identify best process performance
- minimize sample size to save time and money
- use an experimental design which guards against interactions among factors.

Small, efficient statistical experimental designs and a statistical problem solving strategy are the key elements of the problem solving approach.

Experimental factors

After examining the production process, six process variables or factors were identified to include in a series of statistically designed experiments. Initial "best guesses" for the factors were based on previous production experience.

The production process begins with various treatments of the mice to encourage cell growth. A schematic of the *in vivo* production of monoclonal antibodies is presented in Figure 2.1.

Some cell lines do not grow well in fully immunocompetent animals so the mice are given an initial radiation treatment to provide partial immunosuppression. A nonlethal dose of Co^{60} gamma radiation may be used for this purpose. The dosage level of immunosuppressive radiation (RadDos) is the first process variable (experimental factor) to include in the experiment. Two dose levels were proposed; namely, 250 and 500 rads.

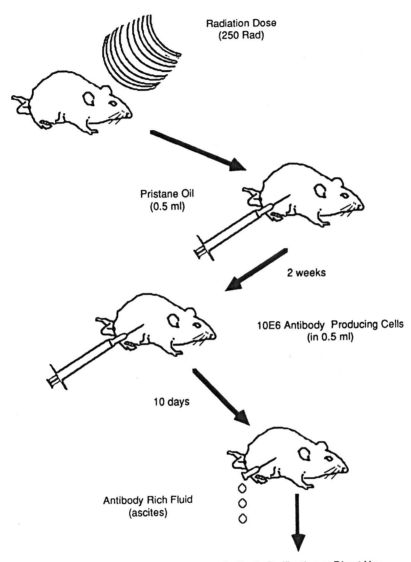

Figure 2.1 Mice may be used to produce an antibody rich fluid called ascites. A general production sequence for this process begins with each animal receiving an immunosuppressive treatment. In this particular case, the immunosuppressive sequence includes irradiation and injection of Pristane oil. Next each animal is injected with antibody secreting cells. After a growth period, the antibody rich fluid, or ascites, can be harvested. (Adrion, *et al*, 1984.)

Additional immunosuppression is obtained by injection of a clear oil called Pristane. Both the amount (VolPrs) and timing of this injection (Prime1) may affect the monoclonal antibody yield. For the experiment, injections of 0.1 and 0.5 ml were used. The elapsed time between the Pristane injection and cell inoculation may vary from a few days to several weeks. Intervals of one and three weeks were used for the first experiment.

In order to start the production of ascites, the mice are inoculated with the antibody secreting cells which were previously cultured *in vitro*. Both the number of cells injected (CelNum) and the growth state of the cells (Growth) are thought to affect yields. Investigators commonly use either 10E6 or 10E7 cells for injection. Log stage of growth and stationary, saturated cultures were proposed for study. (When the values of a variable represent categories, such as "log stage" for the variable Growth, the variable is said to be *qualitative*. A variable which represents a measurement, such as 10E6 cells for CelNum, is called *quantitative*.)

Finally, a second injection of Pristane may sometimes be given immediately before injection of the antibody secreting cells (Prime2). It is unknown whether or not this second injection immediately before tumor inoculation is helpful so experiments were conducted both with and without a second injection.

In order to get enough ascites to provide a reliable assay result, five mice were used for each experimental condition. At the beginning of the study, individual mice were randomly assigned to groups of five. The five mice in each set were housed together and given identical treatment. The ascites collected from the individual mice in each group of 5 was pooled to provide a sufficient volume for a reliable assay of antibody yield. The response measured was antibody titer adjusted for volume (TtrVol), which is proportional to the number of monoclonal antibody molecules produced.

A number of other process variables were of interest but were more difficult and expensive to investigate. In particular, six week old, female, Balb/c mice were used in this study. Six week old mice were used because of the prohibitively high upkeep cost while the animals age. Balb/c mice were used in the study because many strains of hybrid mice cannot be purchased in large quantities on demand. Female mice were used because of difficulties in housing males together. In addition, only one type of monoclonal antibody was considered.

Since these factors were not included in the study, we don't know how changing them would affect the outcome of the study. But if they do change at a later date, the usefulness of these problem solving methods for optimizing the process has been clearly demonstrated.

2.2 A screening experiment

The objective of a screening experiment is to determine which few process variables, out of many candidates, have an important affect on process performance. Designs for screening experiments are introduced in Chapter 3 and used extensively throughout the rest of the book. Complete descriptions of these designs and instructions for their use are included in the Design Digest at the end of this book. The procedure for selecting the best experimental design is covered in Chapter 5. The analysis of screening experiments is presented in Chapter 4. In this chapter we present an overview of how screening experiments are used in statistical problem solving.

Experimental design

The factors to be included in the screening experiment and their settings are given in Table 2.1. Each of the six factors are to be investigated at two levels. Using a design from the Design Digest, we can efficiently investigate these six factors in a sixteen run experiment; that is, by measuring antibody yields for sixteen different combinations of the settings of the process variables.

The experimental design used for the screening experiment is shown in Table 2.2. It is called a fractional factorial design because it is a specially selected subset of the design conditions from the full matrix or full factorial design. (This design is listed in the Design Digest as FF0616 -Fractional Factorial - 6 factors - 16 observations.) It is a time saving design because it requires only a fraction of the experimental conditions of a full factorial design. In partic-

Table 2.1 Variable Names and Levels for First Experiment

Factor Name	Units	Low Level	High Level
RadDos	rads	250	500
Prime1	weeks	1	3
VolPrs	ml	0.1	0.5
CelNum	cells	10E6	10E7
Growth	state	Log	Sat
Prime2	present	No	Yes

Response	Units		
TtrVol	Proportional to number of molecules of monoclonal antibody produced		

Table 2.2 Worksheet for Sixteen Run Screening Experiment

		Experimental Factors					Response
Run*	RadDos	Prime1	VolPrs	CelNum	Growth	Prime2	TtrVol
1.	250	1wk	0.1ml	10E6	Log	No	70
2.	250	1wk	0.1ml	10E7	Sat	No	150
3.	250	1wk	0.5ml	10E6	Sat	Yes	34
4.	250	1wk	0.5ml	10E7	Log	Yes	32
5.	250	3wk	0.1ml	10E6	Sat	Yes	137.5
6.	250	3wk	0.1ml	10E7	Log	Yes	56
7.	250	3wk	0.5ml	10E6	Log	No	123
8.	250	3wk	0.5ml	10E7	Sat	No	225
9.	500	1wk	0.1ml	10E6	Log	Yes	50
10.	500	1wk	0.1ml	10E7	Sat	Yes	2.7
11.	500	1wk	0.5ml	10E6	Sat	No	1.2
12.	500	1wk	0.5ml	10E7	Log	No	12
13.	500	3wk	0.1ml	10E6	Sat	No	90
14.	500	3wk	0.1ml	10E7	Log	No	2.1
15.	500	3wk	0.5ml	10E6	Log	Yes	4
16.	500	3wk	0.5ml	10E7	Sat	Yes	15

* Runs were conducted in randomized order to guard against systematic bias.

ular, a full factorial or matrix design would involve measuring antibody yield at all of the $2^6=64$ different combinations of the values of the six factors.

For comparison, the one-at-a-time method uses fewer conditions than the statistical approach, but requires several weeks of raising and treating mice in order to get results on each of the six factors one at a time. Also, the one-at-a-time approach is not very well adapted to possible interactions among the process variables. The statistical approach, however, consists of two small, efficient experiments; namely, a screening experiment and then an optimization experiment. This iterative approach uses small, economical experiments which are designed to reflect the objectives of different stages in the problem solving process.

For the experimental design worksheet in Table 2.2, the six factors are investigated in sixteen runs. Each run corresponds to a set of values for the factors at which a measurement of the response is to be made. After the experimental factors and responses were specified, the runs were conducted in randomized order to guard against systematic biases. (More details on the use of randomization are included in Chapter 5. Instructions are provided in the Design Digest.) The last column of the table lists the response values which were measured for each run.

This sixteen run design is efficient because it captures the most important information from the full factorial design. That is, these sixteen runs are carefully selected to maximize the information obtained per measurement.

This fractional factorial design also has the desirable property of being balanced; i.e., the effect of a low-to-high change for each factor is measured at both the low and high settings of every other factor. In addition, the calculated effects of the experimental factors are independent of the pairwise interactions among the process variables. The effects of some (but not all) of these interactions may be estimated. Thus, this design provides a careful, well-balanced exploration of the experimental region and a reasonable level of protection against misinterpretations due to interactions among the process variables.

Results of the screening experiment

The results of the experiment are interpreted based on our estimates of how each of the experimental factors affected the response (measure of process performance). The statistical analysis provides estimates of these effects.

In particular, the effect of each factor is the difference in the response value associated with going from the low to the high setting of that factor. An important factor causes a large effect because the process will perform significantly better at one of its two settings. Conversely, an unimportant factor does not result in a large change in process performance and so is associated with a small effect.

In a fractional factorial design, the effect of each factor is the difference between the average of measurements made at the high level of the factor and the average of the measurements made at the low level of that factor. For example, for RadDos, the high settings are runs 9-16. The low settings are runs 1-8. Thus the effect of going from low to high level of RadDos is

$$
\begin{array}{rcccl}
= & 177/8 & - & 827.5/8 & \\
= & 22.1 & - & 103.4 & \\
= & -81.3 & & &
\end{array}
$$

The effects for each of the six factors are shown on the Pareto chart in Figure 2.2.

RadDos has by far the largest effect on ascites yield, and we think that it is clearly a significant factor. The next largest effects are those of Prime2, Growth, and Prime1. We also believe that these factors correspond to important signals. It is not clear whether any of the two-factor interactions are important or not. Unfortunately, the Pareto chart by itself does not allow us to make this distinction. In Chapter 4, we present the statistical methods we use to identify which signals are important.

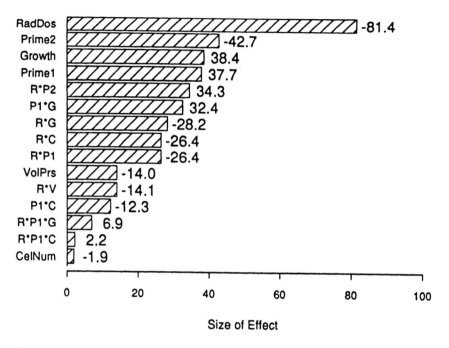

Figure 2.2 A Pareto chart for factor effects displays the magnitude (absolute value) of each estimated effect along the x-axis. The factor names are on the y-axis. The Pareto Principle (effect sparsity assumption) suggests that only a few factors are responsible for most of the changes in the process yield. In this case, RadDos clearly has the biggest effect on ascites yield. Three factors, Prime2, Growth and Prime1, also seem to be important factors. In Chapter 4, additional statistical tools are presented which can be used to determine which additional effects are important.

In order to better interpret the effect of RadDos, we plot the values of Ttr-Vol at the low and high levels of RadDos as in Figure 2.3. We see that much better ascites yields are obtained at the low (250 rads) level than at the high level (500 rads). Correspondingly, the average value of TtrVol at RadDos=250 will be larger than the average of TtrVol at RadDos=500. Thus, RadDos has a large negative effect. We will consider it at lower levels in our next experiment.

Because Growth is a qualitative factor and has a possibly significant positive effect, we should use the high (saturated) level in our next experiment. Prime2 is also a qualitative factor. Its effect is negative which indicates that we should not use a second priming in our second experiment. The other possibly significant factor is Prime1, which is quantitative, so we want to investi-

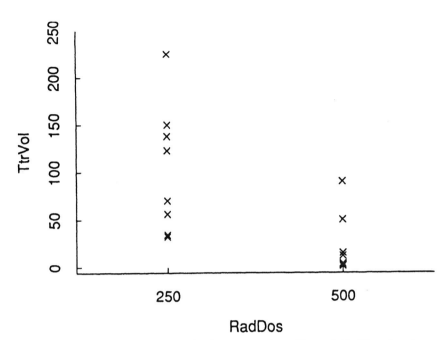

Figure 2.3 This plot illustrates the large negative effect of RadDos by showing that the values of TtrVol are, on the average, larger at the lower level of RadDos (250) than at its higher level (500).

gate it further in our second experiment. Since its coefficient is positive, we will consider it at slightly higher levels in our next experiment. The other two factors, CelNum and VolPrs, do not appear to be important, so they can be set at economical (low) levels of 10E6 and 0.1 ml, respectively.

In order to better understand the results of this experiment, we consider one final graphical display. Figure 2.4 shows "cube plots" for the low and high levels of RadDos. Values of TtrVol are shown at the vertices of the cube. These values correspond to the combinations of high and low settings for the three factors Prime1, Prime2 and Growth. (Cube plots are discussed further in Chapter 4. See also Box, Hunter, and Hunter, 1978.)

Figure 2.4 shows that the best process yields are at the lower level of RadDos (the lower cube corresponds to RadDos=250). By examining the lower cube further, we can understand the important effects of the factors Growth, Prime2, and Prime1. Thus, on the lower cube, we see that saturated growth (back face of the cube) does better than log stage growth (front face of the cube). Better results are also seen on the bottom face (Prime2=no) than on

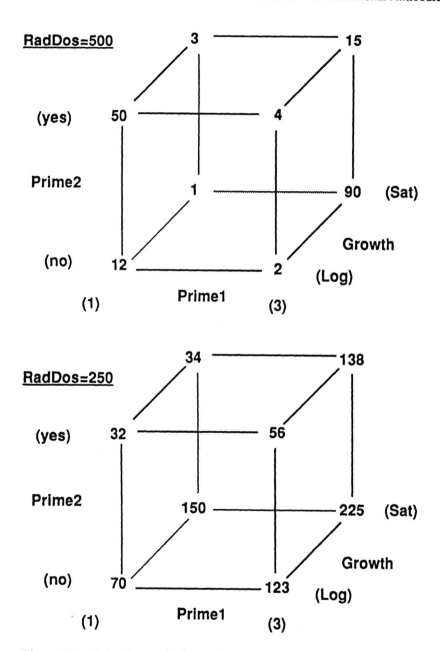

Figure 2.4 Cube plots at high and low levels of RadDos. The best yields are found on the lower cube which corresponds to RadDos=250 rads.

the top face (Prime2=yes). Improvements in process performance are also seen going from the left face of the cube (Prime1=1) to the right face of the cube (Prime1=3). Therefore, Growth has a positive effect (high level is better), Prime2 has a negative effect (low level is better), and Prime1 has a positive effect (high level is better).

Based on the systematic changes noted above, the best value of TtrVol should be at the low level of RadDos, the high level of Prime1, the low level of Prime2 and the high level of Growth. The yield for this set of conditions, 225 units, is in fact the best observed in this experiment. Thus, it makes sense to investigate this promising set of conditions in a subsequent experiment.

Conclusions

The results of the screening experiment indicate that four of the experimental factors are important; namely, RadDos, Prime2, Growth, and Prime1. The two factors Prime2 and Growth can be set at the better of their two values, Prime2=no and Growth=Sat. Further adjustments can be made to the settings of RadDos and Prime1 so they should be included in the next experiment at levels which are closer to their optimal settings. The remaining two factors, CelNum and VolPrs, do not appear to have an important effect on ascites yield so we can set them at economical levels. We are now ready to conduct another small experiment.

2.3   An optimization experiment

In Chapter 1, we discussed the use of an iterative problem solving strategy. One of the advantages of this strategy is that each experiment can be specifically designed to meet the current objectives of the problem solver. In particular, one or more screening experiments may be required to reduce the scope of the problem so that an optimization experiment is feasible. In our example, we went from six to two experimental factors as a result of the screening experiment. Now we are ready to design an optimization experiment.

Objectives

The factors radiation dose (RadDos) and number of days between first priming and inoculation (Prime1) survived the initial screening for important variables. Now we want to develop a mathematical model which can be used to predict the ascites yield (TtrVol) as a function of RadDos and Prime1. This mathematical model provides a prediction of the "best" settings of RadDos and Prime1 which should produce the optimum value of TtrVol.

Optimization experiments fall in the class of response surface methods. In the simple case of two factors and one response, the mathematical model for

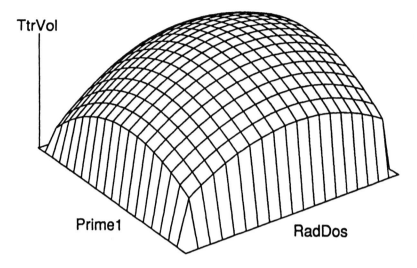

Figure 2.5 This response surface plot is a three dimensional representation of the relationship between the response TtrVol and the experimental factors RadDos and Prime1. In this plot, the response is the z-value (vertical axis), and the factors are the x- and y-axes (horizontal plane). The response surface is generated by a mathematical model which estimates the true functional relationship between the response and the experimental factors.

the response (z) describes a surface over the x-y plane. (This can be generalized to higher dimensions.) This is usually shown in a 3-dimensional surface plot as in Figure 2.5.

If there is an optimum response value somewhere within the experiment region, the surface will be higher in that region. Thus, a curved surface must be considered. Our experimental design includes several levels (five) of each of the two remaining factors in order to estimate the curvature in the response surface.

Experimental design

The design used for the optimization experiment is given in Table 2.3. This is a five-level space filling design known as a central composite design. (This design is listed in the Design Digest as CC0211 - Central Composite - 2 factors - 11 observations.) The center point of the design is repeated three times

Table 2.3 Worksheet for Eleven Run Optimization Experiment

	Factors		Response
Run*	*RadDos*	*Prime1*	*TtrVol*
1.	100	7	207
2.	100	21	257
3.	300	7	306
4.	300	21	570
5.	200	4	315
6.	200	24	154
7.	59	14	100
8.	341	14	513
9.	200	14	630
10.	200	14	528
11.	200	14	609

* Runs are conducted in randomized order to guard against systematic bias.

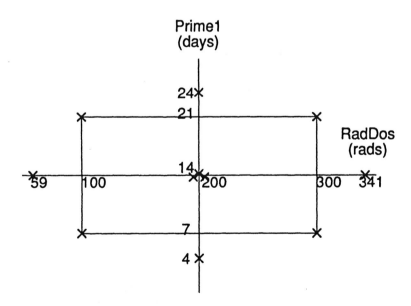

Figure 2.6 Layout of points for the space-filling central composite design

in order to allow a better estimate of the experimental error and to provide extra information about the yields in the interior of the experimental region.

Complete information on all of the experimental designs which we commonly use for optimization experiments is provided in the Design Digest. For more information on optimization designs see Box and Draper (1987), Box, Hunter, and Hunter (1978), Khuri and Cornell (1987) and Myers (1976).

The experimental ranges shown in Table 2.3 reflect the results of the initial screening experiment. Since lower radiation levels were best, the new range for RadDos covers values slightly higher (300 rads) to quite a bit lower (100 rads) than the best setting so far (250 rads). Due to the undesirable economics of longer priming times, the best setting so far for Prime1 (21 days/3 weeks) is toward the upper end of the new range (4-24 days). All of the other factors from the first experiment were fixed at their best values. By using the results of the screening experiment, we hope to avoid a common error in the design of response surface experiments; namely, choosing an experimental region which does not include the optimum settings.

The mathematical model

The model we use to characterize TtrVol as a function of RadDos and Prime1 is a second order (quadratic) polynomial model. It is of the form

$$z = A + B*x + C*y + D*x^2 + E*y^2 + F*x*y + \text{error}$$

The squared terms in x and y represent the curvature in the surface. The linear and two-factor interaction terms are the same as in a fractional factorial experiment. When we conduct an optimization experiment, we hope that the response surface will be curved upward (i.e., hump-shaped as in Figure 2.5) so that the optimum response will be found in the interior of the experimental region.

The statistical analysis of the experimental results provides the following estimated response surface model:

predicted TtrVol =   -608.4 + 5.236*RadDos + 77.0*Prime1
                       - 0.01265*RadDos^2 - 3.243*Prime1^2
                       + .07643*RadDos*Prime1

This model generates a response surface plot for TtrVol and is used for numerical optimization and to generate Figure 2.5. The hump-back shaped of this surface indicates that within the experimental region there are settings for RadDos and Prime1 which produce an optimal response for TtrVol.

Optimization results

Using a numerical optimization program (XSTAT), we determine that the predicted best settings are RadDos = 252 rads and Prime1 = 14.8 days. The predicted value of TtrVol at these settings is 622 units. This agrees well with the response surface in Figure 2.5.

A numerical optimization program is used because it is difficult to read exact values from a graph. With more factors or more than one response, such programs become essential. Two commercially available programs for this purpose are COED/RSM, which is available through the time-sharing service Compuserve, and XSTAT, which is a commercial PC program.

In small problems such as this, good results can also be obtained from applying principles of calculus or even just looking at the response surface plots. See for example Box, Hunter and Hunter (1978) or Khuri and Cornell (1987).

Further evaluation of the optimization results can be made using a contour

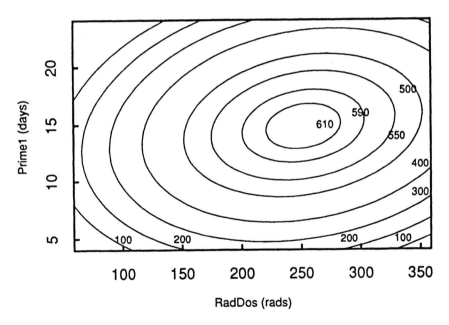

Figure 2.7 This contour plot for TtrVol is very similar to a topographical map used by hikers. The display consists of lines of equal value of the response. The patterns of these contours show how the response changes as a function of the two factors. If we are trying to maximize the response, we are looking for a "bull's eye" pattern which represents a "hill top".

plot (Figure 2.7). The contour plot shows contours of equal values of the response. It is very similar to a topographical map which shows contours of equal surface elevation. The "bulls-eye" pattern in Figure 2.7 is typical of the hump-shaped response surface in Figure 2.5.

One interesting and important feature that can be seen in the contour plot is that the values of TtrVol do not fall off steeply if RadDos and Prime1 change slightly from their best values. This is good because the optimal solution will be robust to slight errors or variability in the experimental factors. This is a very desirable property for an optimized process.

2.4   A verification experiment

The final step in the statistical problem solving process is to verify that the predicted solution works in practice. In order to accomplish this, a production run at the indicated settings was made. The production run produced 22 mg of monoclonal antibody per animal corresponding to a TtrVol of 602 (622 had been predicted). Previous production yields were 11 mg of monoclonal antibody per animal. Thus, a 100% increase in production yield was documented.

2.5   Summary

A screening experiment was used to evaluate six process variables in sixteen different experimental conditions. Each experimental condition corresponds to a particular set of values of the six factors. This screening experiment identified important factors, provided more information about the best settings of the qualitative factors, and pointed us in the direction of optimum settings for the quantitative factors.

The analysis of the data suggested that two experimental factors should be studied further in an optimization study. Two of the other six factors were qualitative factors whose best settings were determined in the screening experiment. The remaining two factors were found to have no significant effect on antibody yield and so were set at economical levels.

The second experiment used eleven experimental conditions as the basis for an optimization study. This data set was used to build a mathematical model which approximated ascites yield. Settings which were predicted to produce the optimum yield were determined for the remaining two experimental factors. Finally the predicted yield at the best settings were verified in a production run.

Thus, a series of small, carefully designed experiments were used to solve a complex biological problem. Each experiment resulted in a small data set which provided information about several process variables. The information gained at each step was used to design the next experiment.

This problem was approached in a way characteristic of both good science and good statistical problem solving. At each step,

- well defined questions were asked,
- clear signal designs were used,
- information-rich data was collected, and
- statistical graphics and analysis were used to provide answers.

After each experiment the results were evaluated and used to plan the next stage. Finally, a well developed strategy was used to both facilitate and monitor progress toward the solution of the overall problem. The end result was a 100% increase in process yield.

References and Bibliography

Adrion, R. F., G. R. Siebert, C. J. Weck, D. Yen, and A. R. Manson (1984). Optimization of *in vivo* monoclonal antibody production using computer-assisted experimental design. *Proceedings of the First Carolina Biomedical Engineering Conference*, North Carolina Biotechnology Center, P. O. Box 12235, Research Triangle Park, NC 27709, 125-144.

Box, G. E. P. and N. R. Draper (1969). *Evolutionary Operation.* New York: Wiley.

Box, G. E. P. and N. R. Draper (1987). *Empirical Model-Building and Response Surfaces.* New York: Wiley.

Box, G. E. P., W. G. Hunter, and J. S. Hunter (1978). *Statistics for Experimenters: An Introduction to Design, Data Analysis, and Model Building.* New York: Wiley.

Cleveland, W. S. (1985). *The Elements of Graphing Data.* Monterey, CA: Wadsworth.

COED/RSM. CompuServe, Inc., 5000 Arlington Centre Blvd., Columbus, OH 43220.

Cornell, J. A. (1980). *Experiments with Mixtures: Designs, Models and the Analysis of Mixture Data.* New York: Wiley.

Khuri, A. I., and J. A. Cornell (1987). *Response Surfaces: Designs and Analyses.* New York: Marcel Dekker, Inc. and Milwaukee: ASQC Quality Press.

Myers, R. H. (1976). *Response Surface Methodology.* Ann Arbor, Mich.: Edwards Brothers (distributors).

XSTAT. Wiley Professional Software, John Wiley & Sons, Inc., 605 Third Avenue, New York, NY 10158.

Chapter 3

CLEAR SIGNAL DESIGNS

A well-designed experiment uses the smallest number of measurements to get the greatest amount of information; that is, it is small and efficient. In addition, a well-designed experiment should allow us to clearly separate signals from the noise so that the data can identify the important factors. We call such designs "clear signal" designs. We have developed a catalog (the Design Digest -- located at the end of this book) of these powerful and efficient designs. In this chapter we explain the properties of these designs and how to use them. For the purposes of presenting these designs, we use a sandwich enzyme-immunoassay example.

3.1   Example: Sandwich enzyme-immunoassay

Enzyme-immunoassay has become an important tool in the determination of serum proteins and hormone levels, therapeutic and illicit drug levels, and viral and bacterial antigens (Maggio, 1980). In many applications it has replaced the use of radioimmunoassay because of its lower cost, nonisotopic nature (absence of radioactive materials), sensitivity, ease of use, and adaptability for automation.

The sandwich assay is a complex biological process. Although the kinetics of such an assay can be modeled (see, for example, Wellington, 1980), there is no theoretical model for the interaction of all the different components of the

37

assay. Consequently, the configuration of such an assay provides a good example for the application of statistical problem solving. In this section, we describe the sandwich assay technique and then use it as an example.

Description of assay technique

The sandwich technique is a widely used enzyme-immunoassay. It is classified as a noncompetitive, solid-phase, direct-binding assay (Butler, 1980). An illustration of the sandwich assay is given in Figure 3.1. In this assay, the antigen in the sample binds with antibody which has been bound to a solid support. The resulting sample is then washed and reacted with an enzyme-antibody conjugate. After a final wash, the enzyme is reacted with a substrate and the resulting product is measured. (Maggio, 1980).

If the sample does not contain the targeted antigen, then the enzyme-antibody conjugate is washed away in the final wash (Step 6). Conversely, if antigen is present in the sample, the enzyme-antibody conjugate is present at Step 7 in proportion to the amount of antigen in the sample.

At the last stage of the assay, the enzyme and substrate react together to form a measurable product, for example a color change. The amount of product is proportional to the amount of antibody-enzyme conjugate bound in Step 5. Thus the reading or measured product should be proportional to the antigen concentration in the sample.

Cause and effect diagram

A good tool for identifying the possible process variables affecting the performance of a sandwich assay is an Ishikawa diagram (Ishikawa, 1976). This diagram is also sometimes called a "cause and effect" diagram or a "fishbone" chart. Its purpose is to stimulate our thinking about possible factors affecting process performance.

In Figure 3.2 we have shown the primary skeleton of the cause and effect diagram for a sandwich assay. This skeleton shows the main factors affecting process performance. Further details could be added to the diagram as necessary; for example, details of the components of the buffers. This diagram is helpful in selecting factors for an experimental design.

Enzyme-antibody conjugate incubation problem

As an example, suppose that a researcher wants to improve the performance of a sandwich assay by optimizing the incubation of the enzyme-antibody conjugate (Step 5 in Figure 3.1). Process performance is measured by the reading obtained when a sample of known concentration is assayed, and it is desired to maximize this reading.

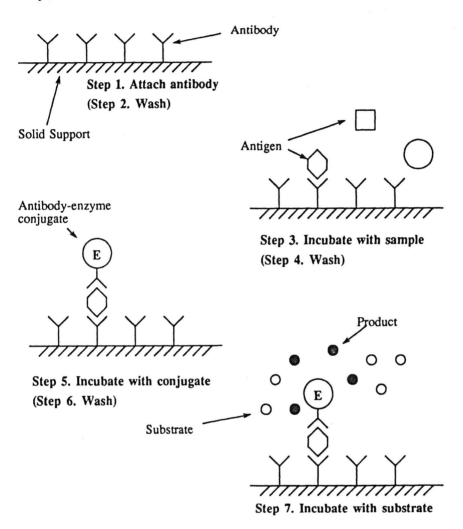

Step 1. Attach antibody
(Step 2. Wash)

Solid Support

Antibody

Antigen

Step 3. Incubate with sample
(Step 4. Wash)

Antibody-enzyme
conjugate

Step 5. Incubate with conjugate
(Step 6. Wash)

Substrate

Product

Step 7. Incubate with substrate

Figure 3.1 A sandwich enzyme-immunoassay generally includes seven steps. Antibody is first bound to the solid support which is generally a plastic tube or microtiter plate (Step 1). Then any unbound antibody is washed off (Step 2). Next a sample containing antigen is added and incubated (Step 3). Following another wash step (Step 4), the antibody-enzyme conjugate is added and incubated (Step 5). After the final wash (Step 6), the substrate is added, incubated, and the resulting product is measured (Step 7). (Clark and Engvall, 1980.)

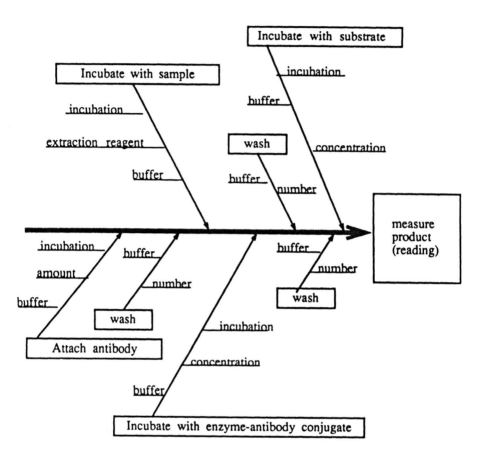

Figure 3.2 An Ishikawa diagram or cause and effect diagram is useful in identifying factors which affect process performance. The first step in constructing such a chart is to show at the far right the process performance characteristic we are measuring along with the heavy arrow. Then the main factors affecting process performance are shown as branch arrows affecting the main process arrow. Additional levels of detail can be added as subarrows until the chart shows all of the factors affecting process performance.

Reviewing Figure 3.2, we see that the process variables involved in incubation of the enzyme-antibody conjugate are the incubation conditions and the buffer characteristics. The primary incubation conditions are the temperature at which the sample is stored and the time it is held. Although there are many factors which make up the buffer characteristics, for this example, we consider only the ionic strength of the buffer.

Let us suppose further that we wish to investigate each of these factors at two levels; namely, incubation time at 20 minutes and 30 minutes, incubation temperature at 27°C and 32°C, and ionic strength at 1.0 mM and 2.0 mM. There are eight combinations of the two levels of these three factors (2x2x2). These eight experimental incubation conditions are shown in Table 3.1. (The values of the response variable, reading, are hypothetical.) This is a full factorial design. We discuss its properties in the next section.

3.2   Two-level factorial designs

Screening experiments are used to identify important factors and to suggest changes in their settings which can improve process performance. The designs which we use most frequently for screening experiments are two-level designs; that is, each factor is evaluated at a "low" setting and at a "high" setting. We are interested in two-level designs because of their ease of interpretation and their effectiveness.

A full factorial design investigates all of the possible combinations of values of each experimental factor. Full factorial designs are examples of "clear signal" designs. A full factorial design allows the independent estimation of the signals associated with each factor and with each combination of factors (interactions). Full factorial designs can also provide good estimates of experimental error or noise.

Table 3.1 Worksheet for Eight Run Factorial Design

Run	Time	Temp	Ionic St.	Reading
1	20 min.	27°C	1.0 mM	1.27
2	20 min.	27°C	2.0 mM	1.18
3	20 min.	32°C	1.0 mM	1.12
4	20 min.	32°C	2.0 mM	0.99
5	30 min.	27°C	1.0 mM	1.36
6	30 min.	27°C	2.0 mM	1.24
7	30 min.	32°C	1.0 mM	1.78
8	30 min.	32°C	2.0 mM	1.69

A two-level full factorial design for three experimental factors would include all 2*2*2=8 combinations of high and low settings of the three factors. A four factor, two-level, full factorial would include $2^4 = 16$ combinations of high and low factor settings. In general, a k-factor, two-level, full factorial design includes 2^k combinations of factor settings.

Each of these combinations of factor levels is called a run. The measurement of process performance (response value) associated with each run is called an observation. Thus, the three-factor, two-level factorial design has 8 runs. The resulting data set has 8 observations. The number of runs or observations, interchangeably, is referred to as the sample size.

Effects of the experimental factors

The signal associated with each experimental factor is called its main effect. For factorial designs, a simple method can be used to calculate an estimate for each of the main effects. The statistical analysis then provides the basis for deciding which of these main effects correspond to large signals and which fall below the noise level.

An estimate of a main effect is obtained by evaluating the difference in process performance caused by changing from the low to the high level of the corresponding factor. The process performance is measured by a response variable. Factorial designs have the special property that the estimates of main effects are simply the differences in average response values between the high and low settings of each factor.

For example, the eight run experimental design in Table 3.1 was chosen for the sandwich assay problem. This is a 2^3 factorial design and is listed in the Design Digest under the name FF0308 (Fractional Factorial with 3 factors and 8 runs). We now show how to estimate the main effects for this example.

Begin by constructing the cube plot in Figure 3.3. Since there are only three factors in our example, the values at the vertices are the corresponding response values (in a larger design, the numbers at the vertices would be the averages of the corresponding runs). This plot shows that the three-factor, two-level factorial design can be represented as a cube. Similarly, factorial designs with more factors can be represented as hypercubes (k dimensional analogs of a cube).

Let us first determine the effect of incubation time on the process performance. Because of the special properties of factorial designs, we can separate out the effect of incubation time by averaging over the other factors. First collapse the cube plot in Figure 3.3 to the square plot in Figure 3.4a by averaging over the front and back faces of the cube (that is, average over the two levels of ionic strength). Then collapse the square plot in Figure 3.4a to the line plot in Figure 3.4b by averaging over incubation temperature.

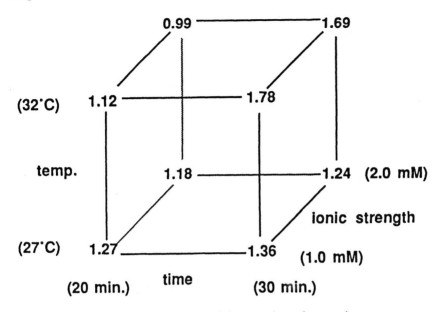

Figure 3.3 A cube plot is constructed for the three factors time, temperature and ionic strength by writing the response value for each run at the appropriate vertex of the cube.

Now the number at the right side of the line plot in Figure 3.4b, 1.52, is the average of the readings for all runs at the high level of incubation time (runs 5-8). The number at the left, 1.14, is the average of all the runs at the low level (runs 1-4). The estimated effect of going from low to high incubation time is the difference of these two averages, namely,

1.52 - 1.14 = 0.38 units

We could estimate the effect of incubation temperature by collapsing the square plot in Figure 3.4a across the levels of incubation time. By doing this we can calculate that the average reading at the high level of temperature (runs 3, 4, 7, and 8) is 1.40 units, and the average reading at the low level of temperature (runs 1, 2, 5, and 6) is 1.26 units. The effect of incubation temperature is the difference in these averages; namely,

1.40 - 1.26 = 0.14 units

Similarly, the estimated effect due to ionic strength is -0.11.

(a) square plot

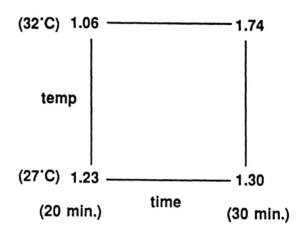

(b) line plot

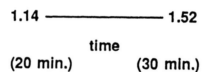

Figure 3.4 The square plot (a) is created by averaging over the third factor, ionic strength, from the cube plot in Figure 3.3. Finally, a line plot (b) is created by averaging over the second factor, temperature.

A Pareto chart for all of the possible effects is given in Figure 3.5. We see that time has the greatest influence on the reading followed by the time*temperature interaction, temperature, and ionic strength. The other effects are negligible. (We show how to estimate the two-factor interaction terms in the next subsection.)

Going from the low to high level of time and temperature resulted in higher readings so these two factors each have positive effects on the response. Going from low to high settings on ionic strength decreased the reading, and therefore, ionic strength has a negative effect. Now we know how to change the factor levels in order to improve process performance.

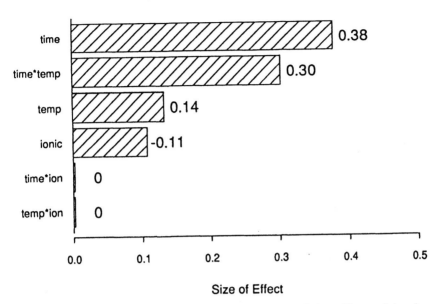

Size of Effect

Figure 3.5 This Pareto chart shows the magnitudes of the effects of incubation time, temperature and ionic strength and their two-factor interactions on the reading of a sandwich assay. The names of the effects are listed on the vertical axis and the absolute values of the effects is shown on the x-axis.

Two-factor Interactions

In the absence of interactions, factors would have an additive effect on the response. Thus, to calculate the joint effect of two factors at their high settings, we would just add together their individual effects. However, if there were a two-factor interaction, then the two factors involved would exert a synergistic (or antagonistic) effect on each other. In this case, the effect of one factor would depend on whether the other factor is at its low or its high value.

One way to identify a two-factor interaction is by looking at an appropriate square plot (Figure 3.4a). We can see that the effect of going from low to high time at low temperature (bottom of square) is much smaller than at the high temperature (top of square). This shows that the effect of time depends on the temperature; that is, these two factors interact with each other.

We can also describe two-factor interactions in terms of a statistical model. Suppose that there are only two factors, say, time and temperature, and there is a single response, reading. Then, if there are no interactions, time and temperature affect the response by the following additive, linear model:

reading $= A + B*time + C*temperature + error.$

In this case, the effects of time and temperature can be graphically represented as two parallel lines (Figure 3.6a). In Figure 3.6a, the upper line represents the effect of going from low to high time at the high temperature. The lower line represents the effect of time at the low temperature.

An interaction changes the relationship as follows:

reading = A + B*time + C*temperature +
                D*time*temperature + error

In this case, the effects of time and temperature are represented by two lines which aren't parallel (Figure 3.6b). The two lines in Figure 3.6b have the same meaning as the lines in Figure 3.6a. However, the effect of time is not the same at the different levels of temperature.

Note that the slope of each of the two lines in Figure 3.6a is equal to the coefficient of time, B. However, in Figure 3.6b, the slope of the line for high temperature is equal to B+D. The slope of the line for low temperature is equal to B-D. This is obtained by substituting, respectively, +1 and -1 values for temperature into the two-factor interaction model above.

The interaction plot for time*temperature in Figure 3.6b can be constructed from the square plot in Figure 3.4a. The line for temp = 27°C is obtained by plotting the points (20,1.23) and (30,1.30) from the bottom edge of the square and then connecting the points. The line for temp = 32°C is obtained by plotting the points on the top edge of the square and connecting the points. The result is shown in Figure 3.6b. Since these two lines are not parallel, there appears to be a strong interaction between time and temperature.

In addition to the interaction plots, the values of the two-factor interaction effects should be estimated. In a factorial design, the estimate of the two-factor interaction effect is the average of the runs in which both factors are at extremes (i.e., high-high and low-low) minus the average of the runs in which the factor levels are mixed (i.e., high-low and low-high). This is the same as taking the difference of the averages of diagonal corners in the square plot of Figure 3.4a. If there is no two-factor interaction this difference is zero (except for experimental error) because the two diagonals have symmetric changes in the factor levels. This is equivalent to the lines in the interaction plot being parallel.

As an example, we calculate the two-factor interaction between time and temperature (time*temperature) for the data in Table 3.1. First, find the average of the runs for which both factors are low or both are high; namely, runs 1 and 2 and runs 7 and 8, respectively, which average to 1.63. Next consider the runs with mixed levels; namely, 3 and 4 (low-high) and 5 and 6 (high-low), which average to 1.18. Then the estimated two-factor interaction between time and temperature is

1.48 - 1.18 = 0.30

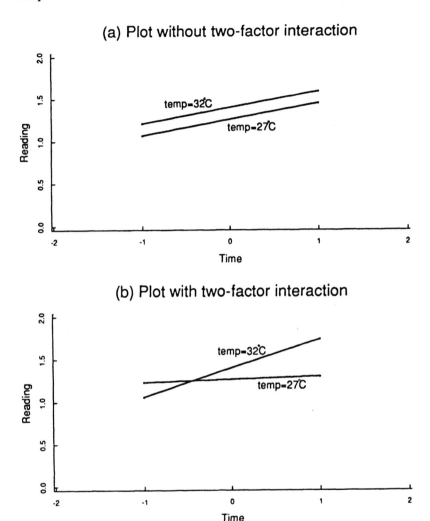

Figure 3.6 An interaction plot graphically shows the presence or absence of a two-factor interaction between two experimental factors. Two lines are plotted. On the y-axis is the response value and the x-axis shows the values for the first of the two factors. One line represents each level of the second factor. When there is no two-factor interaction (a), the lines are parallel. However, in the presence of a two-factor interaction (b), the effect on the response of going from low to high level of the first factor is not the same at each level of the second factor (i.e., the two factors interact). Thus, the two lines are not parallel.

Another way to graphically show the meaning of two-factor interactions is through response surface plots like we used in Chapter 2. Since there are no squared terms, any curvature in the surface is due to the two-factor interaction terms. In Figure 3.7a, we plotted the response surface corresponding to the data in Table 3.1, if we ignored any potential two-factor interactions. In particular, Figure 3.7a was generated for the two factors time and temperature. For the purposes of this plot, we set ionic strength equal to 1.5 mM. Note that the response surface is a plane. Now consider the response surface in Figure 3.7b which includes the time*temperature interaction. Notice that the two-factor interaction causes the surface to twist.

Two-factor interactions can have an important effect on the relationship between the response and the experimental factors. The use of graphical displays such as cube plots, interaction plots and response surface plots can help us better understand the effects of two-factor interactions.

3.3   Fractional factorial designs

Two-level, full factorial designs are very powerful because they provide information about all main effects and two-factor interactions. (In fact, three-factor and higher interactions in a full factorial design can be estimated.) However, the designs which we use most frequently for screening experiments are two-level fractional factorial designs. A design is called a fractional factorial if it is a fraction (subset) of a full factorial. Fractional designs are very efficient because of their smaller sample sizes.

Because of their smaller sample sizes, not all fractional factorial designs are examples of "clear signal" designs in the same way as full factorial designs are. As we go to smaller fractions of the original factorial design, we lose information about possible interactions. For example, a highly fractionated design might clearly distinguish among all of the main effects, but there might be some ambiguity about the two-factor interactions. We balance this loss of information against significant savings in sample size. The information needed to make productive trade-offs between sample size and information about interactions for these designs is provided in the Design Digest.

The use of fractional designs to reduce the sample size required by factorial designs was proposed by Finney (1945). Additional early work in this area was done by Plackett and Burman (1946), Davies and Hay (1950), Daniel (1959), and Box and Hunter (1961a,b). These and similar designs have become commonly used tools for problem solving in industry. Some popular books which discuss fractional factorial designs are Box, Hunter and Hunter (1978), Box and Draper (1987), Diamond (1981), Daniel (1976) and Taguchi (1986).

(a) Response Surface Without Interactions

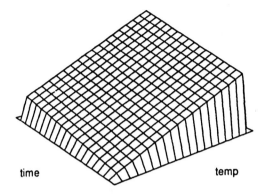

(b) Response Surface With Interactions

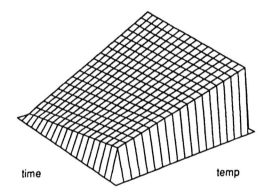

Figure 3.7 The effect of a two-factor interaction can be visualized by considering a response surface plot. In this case, the immunoassay reading is plotted against time and temperature. When the interactions are omitted from the model, a planar surface is obtained (a). When the important two-factor interaction is included in the model, the surface is twisted (b).

2^{k-p} designs

A two-level, fractional factorial design typically represents only one-half, one-fourth, one-eighth, one-sixteenth, etc., of the possible combinations given by the full two-level factorial. If there are k factors, the full factorial has 2^k combinations of factor levels, and a fractional factorial has

$$2^k/2^p = 2^{k-p}$$

combinations. In this scheme, the fraction is $1/2^p$; for example p=1 corresponds to a 1/2 fraction, p=2 corresponds to a 1/4 fraction, etc.

The number of combinations of factor settings included in the design is the sample size for an unreplicated design. (Replication, which increases the sample size by repeating all of the combinations in the design one or more times, will be discussed in Chapter 5.) For example, a full factorial design with 3 factors would have a sample size of $2^3=8$ combinations of high and low settings. Similarly full factorials for 4, 5, and 6 factors have sample sizes of 16, 32, and 64. However, if a design with 4 factors had a sample size of 8, then it would be a one-half fraction of the full factorial; that is, a 2^{4-1} design. Similarly, a 6 factor experiment with a sample size of 16 would be a one-fourth fractional factorial; that is, 2^{6-2}. Usually it is possible to use small fractions and still get good information.

Confounding

In a full factorial design all possible interactions among factors can be estimated. These include two-factor and higher order interactions. However, in a fractional factorial design, we usually give up the ability to estimate some of the higher order interactions in order to be able to use a smaller sample size. How we give up information on interactions is that they become confounded with other terms in the model so that we cannot get independent estimates for them. (See Chapter 5 for more details.)

Resolution

Some fractional factorial designs allow all main effects and two factor interactions to be independently estimated; that is, without confounding. A second type of fractional factorial design allows confounding among two-factor interactions but does not allow any two-factor interaction to be confounded with a main effect. Finally, some fractional factorial designs allow main effects to be confounded with two-factor interactions but do not allow main effects to be confounded with each other. We call these classes of confounding Resolution V, Resolution IV, and Resolution III, respectively.

Although a Resolution V design doesn't allow any confounding among main effects and two factor interactions, it does allow confounding with three-factor interactions. Designs with resolutions higher than V, prevent confounding among successively higher order interaction terms. Usually we are only concerned with up to Resolution V designs because, in practice, we seldom if ever find three-factor interactions which are both important and interpretable.

For fractional factorial designs, a Resolution V design has more information about its factors and their interactions than either Resolution IV or III designs. Similarly a Resolution IV design has more information than a Resolution III design. When selecting an experimental design, we generally want to choose the design with the highest resolution. That is we use Resolution V whenever possible and Resolutions IV and III only when they are specifically appropriate. We never use a design which is worse than Resolution III.

3.4   Clear signal designs

We call a Resolution V (or higher) design a clear signal design because it provides independent estimates of the effects of all of the experimental factors and their two-factor interactions. Thus, signals associated with any of these effects are clearly separate from each other.

It is generally advantageous to use clear signal designs because they minimize the risk of getting ambiguous results. For example, consider an experimental design for an immunoassay system which includes the following factors: time, temperature, pH, ionic strength, and gelatin concentration. A Resolution III design might confound the time*temperature interaction with the effect of pH. Then we couldn't be certain whether a signal was really due to the time*temperature interaction or to pH. Similarly, in a Resolution IV design, we might find a signal which might be attributed to more than one two-factor interaction.

Clear signal designs are usually large enough to estimate all of the effects of interest and also to provide a good estimate of the size of the experimental error. This allows a better determination of the noise level which we use as a yardstick to decide which effects are large enough to be important. As a consequence, we can have more confidence in our conclusions. In Chapter 4, we present this statistical analysis.

We limit the discussion in the rest of this chapter and in Chapter 4 to clear signal (Resolution V) designs. Designs of lower resolution and designs which are not strictly two-level designs are discussed in Chapter 5.

3.5   Selecting an experimental design

The first step in selecting an experimental design is to determine how many factors are to be investigated. Next we consider the need for a clear signal design. Then, we need to decide how big an experiment we are willing to do. With this information in hand, we can try to match our proposed experiment to a design from the Design Digest. If an exact match can't be found, then we must modify our proposed experiment to fit the "best" design available.

Selection of experimental factors

As part of the process of describing a biological process, we come up with a list of potential variables which might affect process performance. These variables constitute our candidate list for experimental factors.

In making up the list of potential experimental factors, we begin with an exhaustive list. A good tool to use in making up this list is an Ishikawa diagram (Ishikawa, 1976). This diagram is also sometimes called a "cause and effect" diagram or a "fishbone" chart. Its purpose is to stimulate our thinking about possible factors affecting process performance.

An example of an Ishikawa diagram for an sandwich enzyme immunoassay system was presented in Figure 3.2. Constructing this diagram, gives us a good overview of the problem. We might then choose to focus on a narrower aspect of the problem or to conduct a broader investigation. As in the earlier example, we might decide that only incubation time, incubation temperature, and ionic strength of the enzyme-antibody conjugate buffer are to be investigated. On the other hand, we might conduct a larger experiment which also included the pH and carrier protein content of the buffer and the number of washes in the preceding and following steps (Steps 4 and 6 in Figure 3.1). We could conduct an even broader experiment by including factors from other steps in the assay procedure.

In general, once we have come up with a candidate list of factors, we should rank order them according to suspected importance. We may also want to then divide this ordered list into three groups; namely; critical factors, probably important factors, and possibly important factors. This ranking helps later if it turns out that we must balance the number of factors to be investigated against sample size and resolution.

Clear signal designs

Whether or not we decide to use a clear signal design depends on a number of things; namely, how much we know about the possibility of two-factor interactions, the number of factors to be considered, and the sample size.

The designs in the Design Digest which are listed as Resolution V designs are the ones which we call clear signal designs. We use these designs when there may be important two-factor interactions. We also must be able to afford a large enough experiment. On the other hand, designs of lower resolution are useful when two-factor interactions are not likely to be important.

For example, the Resolution III designs are often used for screening experiments in which there are many factors, of which only a few are thought to be important. In this case, it is assumed that if there are any two-factor interactions, they are much smaller than the most important main effects. Therefore, we are willing to use a design which allows the two-factor interactions to be confounded with the main effects.

Resolution IV designs are used in intermediate circumstances. For example, we use Resolution IV designs for screening experiments in which there are limitations on sample size and we think that some, but not all, of the two factor interactions may be important. For example, the screening design discussed in Chapter 2 is a Resolution IV design. If there are any signals associated with the two-factor interactions, they are at least clear of the main effects. If there is a need to determine exactly which two-factor interaction caused the signal, we can do a second experiment to decipher them.

Finally, in some cases, we already have a good idea about which two-factor interactions are important. In this case, we pick the smallest design which provides clear signals for all of our main effects and for these predetermined two-factor interactions. This special information allows us to ignore the resolution of the entire design in favor of its specific confounding pattern.

It is quite important to consider whether there may be two-factor interactions, how many interactions there may be, and in some cases, even which two-factor interactions may be important. We prefer to use clear signal designs whenever possible because they give us the best protection against two-factor interactions. However, there are many situations which require screening designs; that is, small designs which handle many factors but largely ignore the possibility of two-factor interactions. We discuss these designs in Chapter 5.

Sample size

By sample size we mean the number of experimental conditions we intend to include in our design. The response variable (or variables) is measured at each experimental condition, and the resulting values are analyzed. The designs in the Design Digest have been specially chosen because they are small and efficient. We most often use 12 and 16 run experimental designs although the Design Digest also contains 8, 20, 24, and 32 run designs.

Some considerations in determining the sample size are how much time it takes to complete the experiment, the expense of the proposed experiment both in time and money, and the availability of raw materials. We generally want to do a small enough experiment so that it can all be completed under homogeneous conditions. On the other hand, the experiment should be large enough so there is a good chance of identifying the important factors (see the discussion on power in Chapter 5).

Homogeneous conditions are especially important for biological experiments since so many unknown, external factors may affect the results of experiments which are not run at the same time and under the same conditions. A few good rules of thumb are

- do a small enough experiment so that it doesn't take longer than a day to complete any one part of it,
- don't use more than one batch/lot of raw materials,
- use only one person to perform all of the work in any one part of the preparation of samples or measurement of results,
- don't use more than one instrument in measuring the results,
- randomize the order in which runs are conducted in order to protect against systematic bias.

These suggestions are intended to provide homogeneous conditions for all of the samples in an experiment. In Chapter 5, we discuss this problem further and describe the use of blocking and randomization to minimize the effects of nonhomogeneous experimental conditions. For now, we assume that the planned sample size is consistent with these recommendations.

Using the Design Digest

Once an experiment has been clearly described, we can consult the Design Digest. The Design Digest has a listing of selected experimental designs and their characteristics. In particular, the Design Digest lists the designs by the number of factors, sample size, and resolution.

Finding a clear signal design is easy. Look for designs which are listed as Resolution V. Designs in the digest are arranged by increasing number of factors. Thus, all of the three factor designs, no matter their resolution, are listed before the four factor designs. For each number of factors, the designs are listed in order of sample size. Because clear signal designs tend to be larger designs, they will usually be listed toward the end of the designs for each number of factors. There is only one two-level, Resolution V, standard fractional factorial design listed for each number of factors. However, in some cases, there may be Resolution V designs which allow one or more of the factors to take on three levels or which are irregular fractions.

The sample sizes required for two-level, clear signal (Resolution V), fractional factorial designs for various numbers of factors are listed in Table 3.2.

The sample sizes required for two-level, clear signal (Resolution V), fractional factorial designs for various numbers of factors are listed in Table 3.2. The clear signal designs for up to 6 factors are included in the Design Digest. We have found the sample sizes for clear signal designs with 7 or more factors to be impractical, and so we have not included these designs in the Design Digest. For reference, they can be found in Box, Hunter and Hunter (1978).

If we find that the sample size required for a clear signal design is too large, then the information in the Design Digest assists us in choosing a compromise design. We discuss this problem in detail in Chapter 5. At that time we also discuss choosing designs which are not clear signal designs.

Examples

In Section 3.2, we discussed a three factor, eight run factorial design for the factors time, temperature and ionic strength (Table 3.1). This is design FF0308 in the Design Digest. It is a Resolution V, clear signal design.

Now suppose that we wish to add the factors pH (at 6.7 and 7.5) and type of carrier protein (casein at 2% or albumin at 1%). This gives us five factors so that if we wish to select a two-level, fractional factorial, clear signal design we need to use the sixteen run design FF0516 (see Table 3.2).

If we also wished to consider the number of washes in Steps 4 and 6 (Figure 3.1), we would then have 7 factors. This design requires 64 runs. If this is too large an experiment, we can choose to add only one of these two factors giving a clear signal design in 32 runs (FF0632), or we could choose to use a design of lower resolution.

Table 3.2 Sample Sizes for Two-Level, Clear Signal (Resolution V)
        Designs

Number of Factors	Sample Size	Design Name
2	4	(not in Design Digest)
3	8	FF0308
4	16	FF0416
5	16	FF0516
6	32	FF0632
7	64	(not in Design Digest)
8	64	(not in Design Digest)
9	128	(not in Design Digest)
10	128	(not in Design Digest)
11	128	(not in Design Digest)

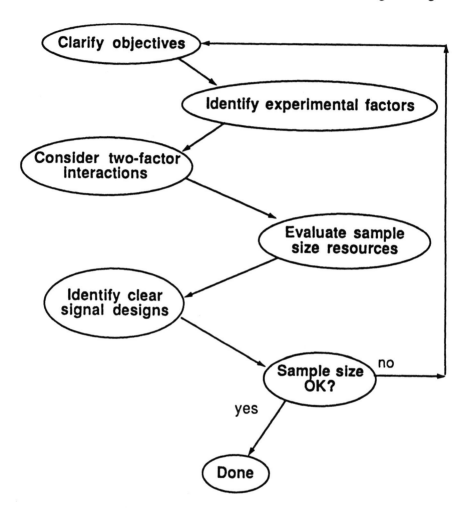

Figure 3.8 The process of selecting a clear signal design includes several steps. First, it is important to clarify the experimental objectives with respect to possible two-factor interactions, then the number of possible factors is determined and sample size limitations are considered. Based on the number of experimental factors, a clear signal design can be chosen from the Design Digest. If this design is not too large, we can proceed with the experiment. Otherwise, we must reevaluate the number of factors to be considered and the importance of getting clear information about the two-factor interactions.

Conclusion

Clear signal designs are advantageous because they provide the best protection against the presence of two-factor interactions. The Design Digest includes two-level, fractional factorial, clear signal designs for from 3 to 6 factors. Once we determine how many factors we want to consider in an experiment, we can look to see how large a sample size is required for a clear signal design. If the sample size is acceptable, we can use the clear signal design confident in its ability to provide good clear estimates of the signals. However, if the sample size is too large, we may have to either reduce the number of factors or use a design with lower resolution. An illustration of this procedure is given in Figure 3.8. Further details on how to select a design are given in Chapter 5.

3.6   Summary

In order to collect data which is information-rich and which reveals the important facts about the process being studied, we must select an appropriate experimental design. In this chapter, we introduced a useful class of two-level designs called clear signal designs. These designs have the attractive property of clearly separating potential signals both from each other and from the background noise (experimental error).

Each particular experimental design specifies which combinations of high and low settings out of all the possible combinations are used in a particular experiment. The experimental data then consist of values of the response at each combination of factor settings (runs). The effects of each factor and two-factor interaction can be estimated by taking simple averages.

In order to choose an appropriate experimental design, we need to understand our objectives, how many factors we need to consider, the possibility of interactions, and how large a sample we can take. We then use this information to select the best design. In this chapter, we only considered two-level, full and fractional factorial, clear signal designs, but there are many other possible designs included in the Design Digest. In Chapter 5, we discuss these designs in more detail.

References and Bibliography

Box, G. E. P. and N. R. Draper (1987). *Empirical Model-Building and Response Surfaces*. New York: Wiley.

Box, G. E. P., and J. S. Hunter (1961a). The 2^{k-p} fractional factorial designs, Part I. *Technometrics*, 3, 311-352.

Box, G. E. P., and J. S. Hunter (1961b). The 2^{k-p} fractional factorial designs, Part II. *Technometrics*, 3, 449-458.

Box, G. E. P., W. G. Hunter, and J. S. Hunter (1978). *Statistics for Experimenters*. New York: Wiley.

Butler, J. E. (1980). Antibody-antigen and antibody-hapten reactions. In *Enzyme-Immunoassay*, E. T. Maggio, editor. Boca Raton, Florida: CRC Press.

Clark, B. R. and E. Engvall (1980). Enzyme-linked immunosorbent assay (ELISA): Theoretical and practical aspects. In *Enzyme-Immunoassay*, E. T. Maggio, editor. Boca Raton, Florida: CRC Press.

Daniel, C. (1959). Use of half-normal plots in interpreting factorial two-level experiments. *Technometrics*, 1, 311-342.

Daniel, C. (1976). *Applications of Statistics to Industrial Experimentation*. New York: Wiley.

Davies, O. L. (1956). *Design and Analysis of Industrial Experiments*, 2nd Edition. New York: Hafner Publishing Company.

Davies, O. L. and W. A. Hay (1950). The construction and uses of fractional factorial designs in industrial research. *Biometrics*, 6, 233-249.

Deming, S. N. and S. L. Morgan (1987). *Experimental Design: a chemometric approach*. New York: Elsevier.

Diamond, W. J. (1981). *Practical Experimental Designs*. Belmont, CA: Lifetime Learning Publications.

Finney, D. J. (1945). The fractional replication of factorial arrangements. *Annals of Eugenics*, 12, 291-301.

Hicks, C. R. (1982). *Fundamental Concepts of Design of Experiments*. New York: Wiley.

Ishikawa, K. (1976). *Guide to Quality Control*. New York: Unipub.

Khuri, A. I., and J. A. Cornell (1987). *Response Surfaces: Designs and Analyses*. New York: Marcel Dekker, Inc. and ASQC Quality Press.

Maggio, E. T. (1980). Enzymes as immunochemical labels. In *Enzyme-Immunoassay*, E. T. Maggio, editor. Boca Raton, Florida: CRC Press.

Montgomery, D. C. (1984). *Design and Analysis of Experiments*, 2nd Edition. New York: Wiley.

Plackett, R. L. and J. P. Burman (1946). The design of optimum multifactorial experiments. *Biometrika*, **33**, 305-325.

Taguchi, G. (1986). *Introduction to Quality Engineering : Designing Quality into Products and Processes.* Tokyo, Japan: Asian Productivity Organization.

Wellington, D. (1980). Mathematical treatments for the analysis of enzyme-immunoassay. In *Enzyme-Immunoassay*, E. T. Maggio, editor. Boca Raton, Florida: CRC Press.

Wheeler, D. (1988). *Understanding Industrial Experimentation.* Knoxville, Tennessee: Statistical Process Controls, Inc.

Chapter 4

SEPARATING SIGNALS FROM THE NOISE

An experimental design leads to a data set which contains information about how the experimental factors affect process performance. This information includes both the signal, i.e., the effects of experimental factors, and noise, i.e., experimental error and measurement error. The statistical analysis provides estimates of the effects associated with each experimental factor and helps us determine which effects or signals are large in comparison to the noise.

4.1   Example: Enzyme-immunoassay stability

In order for a new diagnostic test to reach the market place, the test materials must be shown to be stable over time. In particular, each assay kit must be marked with a shelf-life which gives a date after which the kit should not be used. Although estimation of shelf-life involves specialized statistical studies, fractional factorial designs play a key role in developing stable assay kits.

Stability issues

The individual elements of an enzyme-immunoassay might include a surface coated with trapping antibody or antigen, a sample preparation kit, washing solutions, a preparation containing the enzyme-antibody conjugate and a preparation containing a substrate. When shelf-life is determined for a diagnostic test, the whole kit is stored and then tested as a unit. However, during product development, the stability of each component of the assay is usually considered separately. The stability of any antigen, antibody, and enzyme in the kit is usually critical.

The antigen, antibody, and enzymes are all large proteins or macromolecules. In general, the stability of these macromolecules is not well understood. There is no theoretical model which can be used to develop stable proteins. The typical problem solving approach uses empirical methods to select buffers and formats which enhance stability. Since there are many factors which affect stability and there are possible interactions among these factors, statistical experimental design provides a primary tool for developing stable assay materials.

Degradation mechanisms

Among the primary concerns in the degradation of the antigen, antibody, enzymes, and antibody-enzyme conjugates included in enzyme-immunoassay materials are

- hydrolysis,
- dissociation,
- conformational changes, and
- clipping.

Hydrolysis is a chemical process of decomposition which involves splitting a bond and adding the elements of water. Dissociation involves the loss of binding between the trapping antibody and the solid support or the loss of binding between the antibody and enzyme in the conjugate. Conformational changes in a protein involve a change in its shape or configuration which affects its function. Conformational changes may occur from denaturation or from binding to other molecules. As a consequence of a conformational change, an antibody may no longer recognize the target antigen or an enzyme may no longer react with the substrate to produce the measurable product. Clipping involves the loss of a segment from the protein which results in a loss of functionality.

Each assay component may be subject to more than one of these degradation mechanisms. Many factors may affect the rate at which degradation occurs. These are difficult problems, but they may often be solved by the development of a stability enhancing storage buffer. Some obvious factors about the storage buffer which may affect stability are pH, type of buffer, ionic strength, type of buffer salts, and the addition of various proteins, antimicrobial agents, and other additives.

Problem description

In this example, an experiment was conducted to evaluate the stability of a component of an immunoassay. A number of different experimental buffers were prepared with various additives which may effect stability.

Materials were stored at room temperature for 25 days and then assayed for the amount of the active component. The objective was to minimize the percentage loss in the active component. The loss is expressed as the degradation rate (percent per month). The experimental factors are the pH of the buffer and the presence or absence of four additives; namely, gentamicin (gent), thimerosal (thimer), chelex, and azide.

We think that pH may affect general reactivity including hydrolysis and other types of binding. Gentamicin, thimerosal and azide are antimicrobial agents. It is thought that microbial contamination may lead to clipping of the protein component. These antimicrobial agents can be expected to act in different ways in terms of whether they kill or inhibit the growth of possible microbes in the buffer.

Chelex is a chelating agent which binds calcium and magnesium and other metal ions. Binding with calcium or magnesium may inactivate the protein due to resulting conformational changes. It is also unknown to what extent the presence of ions in the solution may help or inhibit assay performance.

Experimental design

A sixteen run fractional factorial design was selected (FF0516). The experimental design conditions and the responses are presented in Table 4.1. This design is a clear signal (Resolution V) design, which means that

- main effects are not confounded with two-factor interactions and
- two-factor interactions are not confounded among themselves.

Thus, important two-factor interactions cannot not confuse our identification of important main effects. All samples could be prepared under homogeneous conditions so the use of blocking was not necessary.

The design allows the independent estimation of the five main effects; namely, pH, gent, azide, chelex, and thimer. There are also ten possible two-factor interactions among the five factors. Each of these ten two-factor interactions can be independently estimated because this is a Resolution V design.

4.2   Separating signals from the noise

The purpose of the statistical analysis is to determine which experimental factors generate signals which are large in comparison to the noise. We begin by estimating the effects due to each factor and their interactions. Then we use a

Table 4.1 Experimental Worksheet for Immunoassay Stability Study

Run*	pH	Chelex	Azide	Gent	Thimer	Response Rate
1	-1	-1	-1	-1	+1	6.90
2	-1	-1	-1	+1	-1	9.81
3	-1	-1	+1	-1	-1	8.78
4	-1	-1	+1	+1	+1	7.92
5	-1	+1	-1	-1	-1	8.42
6	-1	+1	-1	+1	+1	7.04
7	-1	+1	+1	-1	+1	7.21
8	-1	+1	+1	+1	-1	9.96
9	+1	-1	-1	-1	-1	2.34
10	+1	-1	-1	+1	+1	1.22
11	+1	-1	+1	-1	+1	1.29
12	+1	-1	+1	+1	-1	1.73
13	+1	+1	-1	-1	+1	1.55
14	+1	+1	-1	+1	-1	1.68
15	+1	+1	+1	-1	-1	1.81
16	+1	+1	+1	+1	+1	1.36

* Runs were conducted in randomized order to guard against systematic bias.

number of different graphical and analytical methods to decide which of these estimated effects are large enough to be considered important signals.

Estimates of effects

This design uses all of the possible degrees of freedom (16, or one per observation) to estimate effects. In particular, 5 main effects, 10 two-factor interactions and the overall mean may be estimated. This type of design can be called *effect saturated*. (The term *saturated design*, or *factor saturated* in our terminology, commonly refers to a design which uses all of its available degrees of freedom to estimate main effects.)

In the previous chapter we showed how simple methods can be used to calculate the effects obtained from a fractional factorial design. For example, the effect due to pH is simply the difference in average response between the runs at high pH and those at low pH. These are simple, but tedious, calculations and we leave them to standard computer programs.

The estimates of the five main effects and their ten two-factor interactions are shown on the Pareto chart in Figure 4.1. We see immediately, that pH has by far the greatest effect at -6.63. The next greatest effect is that of thimerosal (thimer) which is -1.26. The third and fourth largest effects are the pH*thimer interaction (0.72) and the pH*gent interaction (-0.55). Most of the remaining effects are quite small.

The Pareto chart provides a useful and practical tool for identifying which estimated effects are the most important. The pattern observed in Figure 4.1 is common of those seen in our practice; namely, there are one or two very large effects, a few intermediate effects, and then a collection of smaller effects. One caution to observe is that the interpretation of the Pareto chart is conditional upon the ranges chosen for the experimental factors. For example, an important factor with its high and low levels set too close together will not look as important on the Pareto chart as a less important factor with a much wider range.

While the Pareto chart makes very clear what the most important factors are, it is less useful in determining the dividing line between the effects which are intermediate in size and the effects which are too small to be of interest. This is where additional statistical methods are useful.

If this design had been replicated (that is, each of the runs had been repeated under independent conditions one or more times), we could estimate the experimental error from the replicates. (See the discussion on replication in Chapter 5 for more information.) Then effects which are large in comparison to the estimated error are judged to be real. However, this unreplicated design provides no independent estimate of the experimental error, so we must use

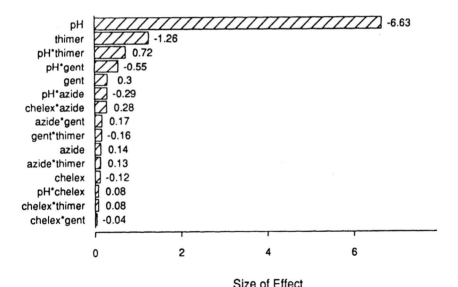

Figure 4.1 Pareto chart for degradation rate.

alternative methods for identifying which effects are real. In this case, graphical methods, such as normal plots and active contrast plots, are very helpful.

Normal plot

A normal plot identifies estimated effects which appear to be large in comparison to the noise. The method is especially useful for unreplicated fractional factorial designs. The plot is constructed so if there were no important effects, all of the points would fall on a straight line. Thus, any effects which are associated with points falling noticeably away from the line may be considered to be real (important signals).

The normal plot for this example is shown in Figure 4.2. If we draw an "eye-fit" line to the data, we are interested in points which fall below the line on the right hand side (large positive effects) and above the line on the left hand side (large negative effects). (See the discussion in Chapter 9 with regard to analytical methods for drawing lines on the normal plot.)

The effect for pH is at the lower left-hand corner of the plot. This effect obviously falls far above the line through the rest of the effects so we believe that it represents a strong signal. Moving over to the right, we find the effect for thimerosal which also seems to fall above an "eye-fit" line and so is probably a signal. It is not clear from the normal plot whether the next two largest effects, pH*thimer and pH*gent are important or not.

Since most commercially available software packages, for example, SAS (1988), Statgraphics (1987), SCA (1987) and SPlus (1987), provide functions to create normal plots, we will not go into detail about how to construct them. Box and Draper (1987) provide a good explanation of how these plots can be constructed by hand in Chapter 4. Box, Hunter and Hunter (1978) discuss normal plots in Chapter 10. Daniel (1976) also provides an interesting discussion of normal plots and other analysis methods for fractional factorial designs. More details are available in Box and Meyer (1986). A similar plot called the half-normal plot was first introduced for analysis of fractional factorial experiments by Cuthbert Daniel (1959).

Active contrast plot

Another graphical display which is useful in identifying important signals in unreplicated designs is the Bayes plot, or active contrast plot. This plot provides a simple graphical display which shows how likely it is that each factor has an important effect on the response.

The active contrast plot gives the (posterior) probability that each effect is real; that is, active. This calculation is based on the assumption of effect sparsity (see Chapter 1) in which we assume that only a few of the estimated effects may be real. Effects which have a probability of 50% or greater of being

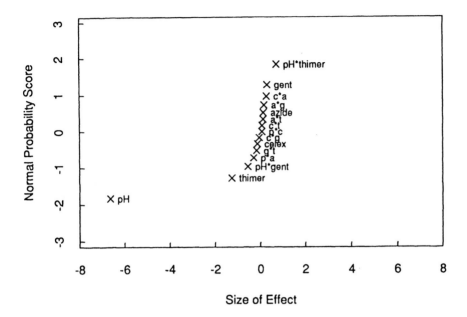

Figure 4.2 Normal probability plot for degradation rate.

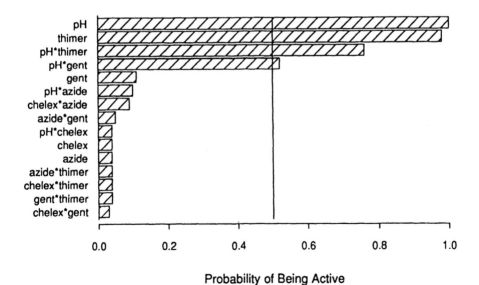

Figure 4.3 Active contrast plot for degradation rate; prior probability=40%, scale factor=20.

active are considered to be important. Box and Meyers (1986) and *Quality and Productivity Improvement Using the SCA System* (1987) provide further details. The SCA (1987) system provides an easy to use program for creating active contrast plots.

The active contrast plot depends on two parameters which must be specified by the user. The first is the prior probability that any particular effect is important (active). Our experience is that a 40% prior probability is a good choice for screening experiments. This means that we expect as many as 40% of the effects to be active. Thus, for 15 effects (5 main effects and 10 two-factor interactions), we are willing to believe in advance that as many as .4*15=6 of them may be important. This is choice places its emphasis on not missing any important effects.

The second parameter of the active contrast plot is the scale factor. This factor specifies, in general, how much larger the estimates of the important effects are than those of the unimportant effects. In contrast to the prior probability, which we use at 40% for all of our data sets, the best value for the scale factor seems to vary from data set to data set. We have adopted the approach of making active contrast plots for several reasonable values of the scale factor (5, 10, 15, and 20). Then we review the active contrast plots along with the Pareto chart and normal plot. Finally, we choose the scale factor which gives the "best" active contrast plot. As in the choice of prior probability, "best" here means that we want to avoid missing any potentially important effects.

The active contrast plot for this data set is shown in Figure 4.3. The effects pH, thimer, pH*thimer (p*t), and pH*gent (p*g) have probabilities of being active which are greater than 0.50 (50%). Note that the active contrast plot is similar to the Pareto chart of Figure 4.1. However, the active contrast plot has made it easier to distinguish between the important and unimportant factors by plotting a probability score instead of the actual sizes of the effects.

The prior probability for Figure 4.3 was 40% and the scale factor was 20. When the scale factor was 5, only pH and thimer were above 0.50. With the scale factor increased to 10, pH*thimer went above 0.50. At 15, pH*gent was close to the 0.50 level. Finally, increasing the scale factor to 20, resulted in pH*gent going above the 0.50 level. In general, we select the smallest scale factor which clearly separates the effects into "important" and "unimportant" sets.

Cube plots

Cube plots provide another useful means for interpreting the results of a fractional factorial experiment. We first introduced them in Chapter 3. We use them here to further investigate the effects which we have identified as being important; namely, pH, thimer, pH*thimer and pH*gent.

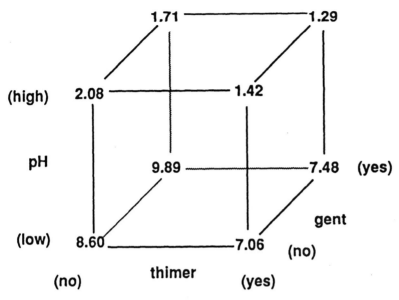

Figure 4.4 Cube plot for pH, thimer, and gent showing degradation rate.

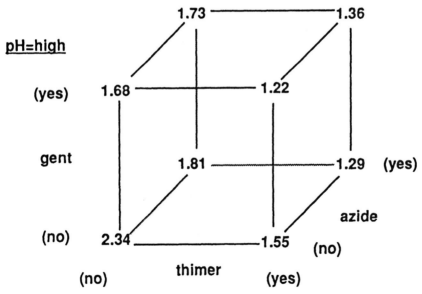

Figure 4.5 Cube plot at high level of pH showing degradation rate.

The first cube plot we consider shows the three variables involved in the important effects (Figure 4.4). The value shown at the vertex is the average of the corresponding runs. We notice immediately that the top face of the cube (high pH) has markedly lower degradation rates. The right face of the cube (thimer=yes) also has lower degradation rates. The effect of the pH*thimer interaction is that the addition of thimer at high pH has a slightly smaller effect than at low pH. The pH*gent interaction is that at low pH better performance (lower degradation) is observed without gentamicin while at high pH better performance is observed with gentamicin. The best vertex of the cube, average degradation rate = 1.29, is for high pH with both thimerosal and gentamicin.

Since pH has such a strong effect on the degradation rate, it is useful to examine a second cube plot which shows results at high pH only (Figure 4.5). In order to decide which variable to add to this plot, we reviewed the Pareto chart in Figure 4.1. We chose to add azide to the cube plot because it was the next largest main effect and because it had several small interactions with the other factors.

The best set of conditions on the cube plot in Figure 4.5 is (for high pH), high thimer, high gent, no azide. If we consider only the right face of the cube (thimer = yes), it appears that we could add either azide (1.29) or gentamicin (1.22) to the buffer but not both (1.36). However, gentamicin has a more important effect than azide both in this cube plot and in the Pareto chart.

Interaction plots

Interaction plots provide a means for better understanding the two-factor interaction pH*thimer and pH*gent. To construct an interaction plot, we calculate the average response at each of the four combinations of high and low settings for the two variables. (Try making a square plot for the two factors first. Refer to Chapter 3.) Usually we put the variable with the larger effect on the x-axis. Then the points with equal settings on the second variable are connected by lines. If there is no interaction, then the two lines should be parallel.

The interaction plot for pH and thimer is shown in Figure 4.6. Since the two lines aren't parallel, we believe that there is an interaction. The pH*thimer interaction is called an inconsistent interaction because it works against the effects of its two factors. However, since the two lines don't cross, we know that the interaction effect is not greater than the main effect of thimer.

When attempting to minimize a response, an inconsistent interaction will have the same sign as the product of the signs of its main effects; in this case sign(pH*thimer) = + and sign(pH)*sign(thimer)=-*-=+. Thus if we attempt to minimize the degradation rate by setting pH and thimer to their high (+1) levels, the pH*thimer interaction works against us by increasing the degradation rate (positive coefficient).

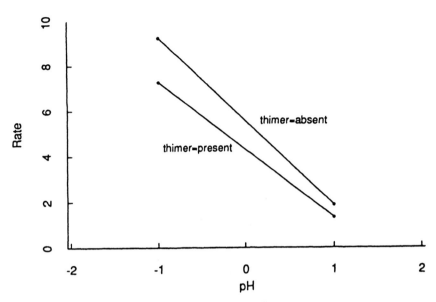

Figure 4.6 Interaction plot for pH and thimerosal.

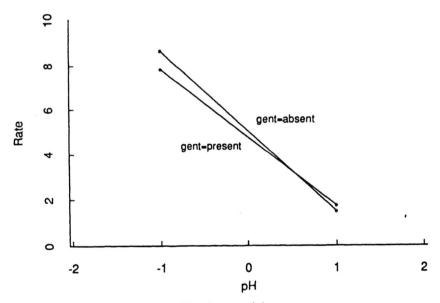

Figure 4.7 Interaction plot for pH and gentamicin.

The reverse situation holds true in identifying an inconsistent interaction when attempting to maximize a response. In this case, an inconsistent interaction would have the opposite sign from the product of the signs of its main effects.

The second interaction plot of interest for pH and gentamicin (Figure 4.7) shows that gentamicin switches its effect between low and high pH. Since, the lines cross, the two-factor interaction is bigger than the main effect of gentamicin. (See also the Pareto chart in Figure 4.1.) Once again, this is an inconsistent interaction. Notice that the effects of gentamicin and its two-factor interaction are smaller than the effects of thimerosal.

Conclusion

The important factors identified by the graphical analysis are pH, thimer, pH*thimer, and pH*gent interaction. Azide and chelex were not important by themselves or through interaction with other factors. Gentamicin was important because of its interaction with pH. These results suggest that in order to improve the stability of this immunoassay component, we should

- use the higher pH,
- add thimerosal, and
- add gentamicin

when formulating a new storage buffer.

4.3   Fitting a statistical model

Now we fit a statistical model containing the important terms identified by the graphical analysis and see how well the model explains the observed variation in the degradation rates.

Fitting the reduced model

The statistical model is called a linear model and is constructed using regression analysis. The model is a mathematical equation which can be used to predict the response. Our analysis so far suggests the following model:

$$\text{rate} = A + B*pH + C*\text{thimer} + D*\text{gent} +$$
$$E*pH*\text{thimer} + F*pH*\text{gent} + \text{error}$$

The coefficients, A through F, represent the true unknown values which we call parameters. Estimates for these parameters are provided by the statistical analysis of the experimental results.

In performing the regression analysis, -1 and +1 are used for the low and high levels of each factor. This allows us to directly compare the coefficients for different factors even if the factors are in different units. Also, the coefficients in the predictive model are then the corresponding effects divided by 2. The effects shown on the Pareto chart (Figure 4.1) can be used to calculate the coefficients (except the intercept term) in the following model:

predicted rate =   4.94 - 3.32*pH - 0.63*thimer +0.15*gent +
0 .36*pH*thimer - 0.28*pH*gent

Since this is an estimated equation and not a theoretical one, we no longer include the error term.

The coefficients can be gotten directly from the Pareto chart because for standard fractional factorial designs, estimates of effects are independent. Designs with this property are said to be orthogonal. For orthogonal designs, the estimated coefficient for any coefficient does not depend on which other terms are included in the model. In Chapter 7, we will study designs which, in order to achieve smaller sample sizes, are slightly nonorthogonal. These nonorthogonal designs have some dependence among estimates of effects.

The predicted values for rate can be calculated for each run by substituting the appropriate values (+1 or - 1) of pH, thimer, and gent into this equation. The absence of other factors from the equation means that we don't believe that the other factors have important effects on the degradation rate.

Estimating the experimental error

The best way to obtain an estimate of the experimental error is by repeating one or more experimental conditions. In many cases, the entire experiment is repeated (replicated) a second time. In some cases, repeated observations are taken only at a single point in the middle of the design space (a center point). The estimate of experimental error is essentially the standard deviation of the repeated observations. More information on center points and replication is presented in Chapter 5 and in the Design Digest.

In an unreplicated fractional factorial design, as in this example, there is no independent estimate of the experimental error. However, only three of the original five experimental factors are now thought to be important. (This is an example of *factor sparsity*, see Box and Meyer, 1986.) Consequently, the sixteen run experiment can be viewed as a twice replicated, 8 run, full factorial design in the three important factors.

By fitting the reduced model described above, we are taking advantage of this implicit replication to estimate the experimental error. In particular, the resulting estimate of the experimental error is 0.331 (See Table 4.2). In practice, this estimate includes, in addition to the experimental error, any additional small effects due to factors and/or effects not included in the model. This estimate will be too large if we mistakenly included one or more important effects in the experimental error. It will be too small if we mistakenly included one or

Table 4.2 Analysis of Variance (ANOVA) Table for the Reduced Model

R-squared 0.994
Adjusted R-Squared 0.991
Estimated experimental error 0.331

Source	SS	DF	MS	F-Ratio
Regression	186.	5	37.2	340.
Residual	1.10	10	0.11	
Adjusted total	187.1	15		

Factor	Coeff.	St.Err.	T-Value	P-Value
Intercept	4.94	0.0827		
pH	-3.32	0.0827	-40.1	<.0001
Thimer	-0.63	0.0827	-7.58	<.0001
Gent	0.15	0.0827	1.83	.10
pH*thimer	0.36	0.0827	4.35	<.01
pH*gent	-0.28	0.0827	-3.34	<.01

more unimportant effects in the model. By using normal plots, active contrast plots, cube plots, and interaction plots, we hope to avoid either of these two mistakes.

4.4 Evaluating the model

The statistical analysis began with the estimation of the effects of each experimental factor and their two-factor interactions. Because the design, FF0516, is an unreplicated design we used several graphical methods to identify which estimated effects are large in comparison to the noise. The next step in the analysis of this data set is to fit a statistical model containing only important effects. This is called a reduced model. In this model, we expect that the effects pH, thimerosal, pH*thimer, and pH*gent will be significantly greater than the noise level. We do not necessarily expect gent to be above the noise level, because we are including it primarily because of its important two-factor interaction with pH.

Definitions of statistical terms

After selecting and fitting the model for predicting the degradation rate, a statistical analysis is performed. The output of this analysis includes the following information:

• An estimated coefficient for each term in the model.

Because we always code the data before conducting the statistical analysis (i.e., -1 = low, +1 = high), we can directly compare the coefficients for the different factors among themselves. However, it is also useful to have a further objective way of judging the relative importance of the different factors.

• The standard error for each coefficient.

This is a measure of the experimental error as it affects each coefficient. It combines the estimate of experimental error with information about the sample size. Because the design is balanced (each factor has the same number of high's and low's), the standard errors are the same for each coefficient.

• A t-value for each coefficient.

The t-value measures how large the coefficient is in relationship to its standard error. Think of this as a "signal-to-noise" type measure. The t-values are obtained by dividing each coefficient by its standard error.

• A p-value for each t-value.

The p-value is the chance of getting a larger t-value (in absolute value) by chance alone. This probability is based on the assumption that the random error associated with the model is normally distributed. A small p-value suggests that the coefficient is a large signal in comparison to the noise because it is too large to have arisen by chance alone. P-values are provided by most statistical analysis software.

Coefficients with small p-values are significantly greater than zero; that is, they identify effects which appear to be truly important. For example, $p < .10$, suggests significance at the .10 level ($\alpha = .10$). This also corresponds to a 90% confidence level for a test of the hypothesis that the coefficient in question is equal to zero. Small p-values are associated with large t-values because they imply that the coefficient is much greater than its standard error.

The values of the estimated coefficients, standard errors, t-values and p-values for our example are given in Table 4.2. The factor pH has by far the largest t-value (-40.1) and seems to be very significant (p<.0001). Thimer and pH*thimer have smaller t-values (-7.58 and 4.35, respectively) but are still highly significant (p<.0001 and p<.01, respectively). The t-value for pH*gent is -3.34 which is also significant at p<.01. As we expected, the t-value for gent (1.83) is not significant (p=.10). Based on these results, we believe that the effects which we selected to included in the reduced model are real.

Interpreting p-values

Caution should be used in interpreting p-values. For a variety of reasons, the p-value given by the statistical analysis may only roughly correspond to a true significance level. This is especially true if we are analyzing an unreplicated design or if the sample size is small. Another reason for caution is that the standard assumptions of independence and normality are seldom fully realized.

The p-value or significance level is a function of sample size so that the same numerical coefficient may be significant in a large sample but not in a small sample. Significance is also a function of the model being used. The same data can give different significance levels if different terms are included in the particular model. This is especially important if we made an error in dropping nonsignificant terms from the model.

Significance may also be a function of the experimental region so that if the high and low settings are too close together or too far apart an important factor may not appear to be significant. Some of the strategies discussed in Chapter 8 are helpful in making good decisions without relying too heavily on p-values or significance levels.

If there is an outlier or unusual value in a data set, it may affect the significance levels of all of the factors, making it difficult to interpret the results. Another problem is that the practical significance of an experimental factor may not coincide with its statistical significance. For example, a small coefficient may be significant if the experimental error is small. Conversely, the importance of a large coefficient may be masked by a large experimental error.

In the scientific literature, p-values $< .05$ are generally considered to be significant. We would like to caution the reader against accepting this as the definitive significance level. For an unreplicated fractional factorial design, important factors should be identified based on all of the information available from the graphical displays, the statistical analysis, practical knowledge of the problem, and the problem solving strategy being used. (See also the strategy discussion in Chapter 8 concerning identifying important factors.) To the extent that p-values are useful for screening experiments, it is probably better to accept higher p-values, say $p < .10$, rather than to take the chance of missing an important factor.

The practical judgement of the experimenter is the final arbiter of the meaning of significant p-values. Use common sense and good graphical techniques in interpreting the statistical analysis. Be aware of the following syndrome:

Statistics ON, Brain OFF

R-squared

The R-squared value provides a measure of how much of the variability in the observed response values can be explained by the experimental factors and their interactions. The R-squared value is always between zero and one. A value of one indicates that the statistical model explains all of the variability in the data. A value of zero indicates that none of the variability in the response can be explained by the experimental factors.

The R-squared value is a standard part of the output from statistical packages. The R-squared value for our example is provided in Table 4.2. Its value is 0.994. Since this value is very close to 1.00, we conclude that the important effects we identified explain most of the variability in the assay reading.

A practical rule of thumb for evaluating the R-squared is that it should be at least 0.75 or greater. Values above 0.90 are considered very good. Most regression tables also provide a F-ratio for the regression model. This F-ratio can be regarded as a test of the assumption that the R-squared is equal to zero. This F-value from Table 4.2 is 340. This is a very significant value (p<.0001).

When expressed as a percent (R-squared*100%), the R-squared is interpreted as the percent of variability in the response explained by the statistical model. Thus, for our example, more than 99% of the variability in the response can be explained by the factors pH, thimerosal, gent and the two-factor interactions with pH.

The R-squared value is also called the coefficient of determination. This is because the R-squared value is calculated from the following relationship:

R-squared = variation explained by model / total variation

Thus, the larger the R-squared, the more accurately the value of the response can be predicted by the model.

Most statistical analysis software packages provide an analysis of variance (ANOVA) table similar to the one shown in Table 4.2. The Regression Sum of Squares is the variation explained by the model and the Adjusted Total Sum of Squares is the total variation (adjusted for the mean). The R-squared value is calculated from Table 4.3 as follows:

R-squared   =   186.0/187.1
               =   0.994

The ANOVA table or regression analysis also includes a value called the "adjusted R-square". The adjusted R-square corrects the R-square value for the sample size and for the number of terms in the model (Draper and Smith, 1981). If there are many terms in the model and the sample size is not very large, the adjusted R-square may be noticeably smaller than the R-square. This should be a caution signal that too many terms are in the model (this may indicate a pooling error). For our example, the adjusted R-square value of 0.991 is quite high.

A good model explains most of the variation in the response. The R-squared is a measure for this criterion. The closer the R-squared value is to 1.00, the stronger the model is and the better it predicts the response. Since the reduced model had a very high R-square value with only a few terms in the model, we feel that we did a good job of finding the factors which have an important effect on assay stability.

Diagnostic plots

There are several additional diagnostic plots to consider when evaluating the goodness of the reduced statistical model. These are the predicted versus observed plot and three residual plots.

The predicted values are calculated from the statistical model. The observed values are the actual values of the response. If the model successfully predicts the responses, then the points on the predicted versus observed plot will lie close to a line with slope of one which passes through the origin.

The predicted versus observed plot for the reduced model (Figure 4.8) meets this criterion. A weak statistical model is suspected if some or all of the points are far away from the line. Any obvious curvature in the pattern of the points in this plot may suggest that a transformation of the response is required. (See Chapter 6.)

The residuals from the model are the differences between the predicted and observed values. In particular,

residual   =   error in prediction
                  =   observed value - predicted value

If the statistical model does a good job of predicting the observed values, then the residuals should be small and should not exhibit any unusual behavior. For example, a single large residual would indicate a possible outlier. If the model overpredicts the response in some areas but underpredicts in other areas, then a plot of the residuals will show corresponding areas of primarily positive or negative residuals. Finally, a pattern of increasing or decreasing spread (variability) in the residuals might indicate the need for a transformation.

The first residual plot which we examine is the plot of residuals versus the predicted values in Figure 4.9. The residual values seem small enough and evenly distributed above and below zero over the range of the data. However, there does appear to be more spread associated with the higher predicted values. This behavior, called heteroscedasticity, suggests a transformation of the response which will be discussed in Chapter 6.

In order for the p-values which we calculated above to be valid, the residuals should be approximately normally distributed. This can be verified in the second residual plot which is a normal probability plot of the residuals (Figure 4.10). This plot is similar to the normal plot used to evaluate the sizes of the estimated effects. As in Figure 4.2, we want the points to fall on a straight line. Any points far away from the line or any deviation from the line (for example, an S-shaped curve instead of a straight line) indicate possible outliers or

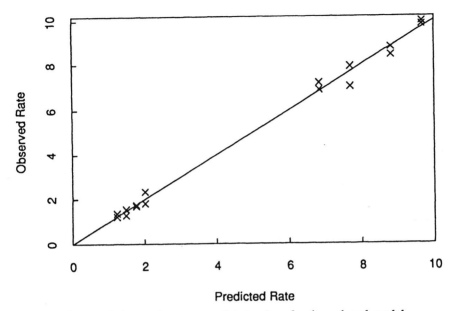

Figure 4.8 Plot of observed versus predicted values for the reduced model.

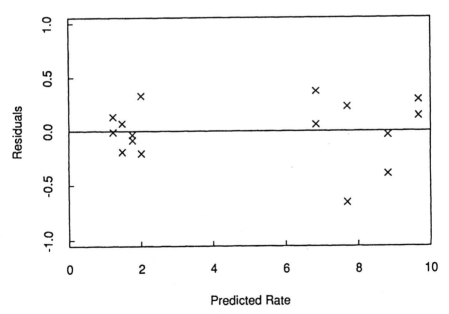

Figure 4.9 Plot of residuals versus predicted values for reduced model.

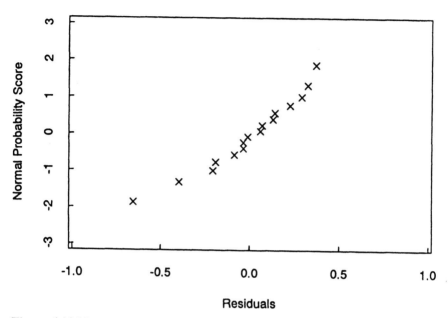

Figure 4.10 Normal plot of residuals from reduced model.

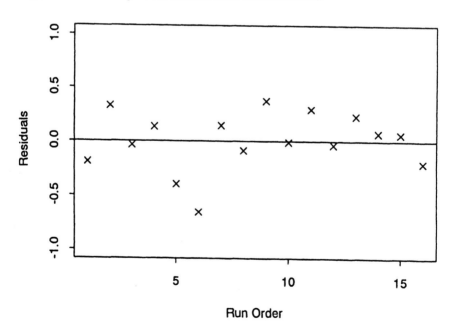

Figure 4.11 Plot of residuals versus run order for reduced model.

some other problem with the model. See Carroll and Ruppert (1988), Box, Hunter and Hunter (1978), Box and Draper (1987) and Daniel (1976) for further discussion of normal plots.

The final residual plot we consider shows the residuals versus run order (Figure 4.11). (The order in which the runs were conducted was as follows: 11, 9, 3, 16, 5, 6, 2, 14, 7, 10, 8, 12, 4, 13, 1, and 15.) Patterns in this plot may indicate some systematic bias in the data. For example, it appears that the 5th and 6th observations (which correspond to runs 5 and 6 in Table 4.1) had unusually large negative residuals. It also appears that most of the residuals alternate between positive and negative values. The presence of these patterns suggests that we should go back and reexamine the laboratory notebooks and talk to the technicians who ran the experiment. In this particular case, no assignable causes could be found. Since the observed patterns do not appear to be very serious, the runs were conducted in a randomized order and there are no assignable causes, we do not believe that any systematic bias has influenced our conclusions.

Conclusions

In this section, we fitted a reduced statistical model which contained the important terms identified from the graphical analysis of the estimated effects. The reduced model was examined based on R-squared values, t-values, and p-values and found to be quite satisfactory. Several diagnostic plots were constructed to further evaluate how well the reduced statistical model fit the observed degradation values. The plot of residuals versus predicted values suggests that a transformation of the response would improve the fit of the model (see Chapter 6). No other serious reservations were suggested by the diagnostic plots.

4.5   Summary

A typical experimental design includes several factors and their two-factor interactions. Any of these effects may be important, but the assumption of effect sparsity suggests that only a few of the effects are responsible for the biggest improvements in process performance. The objective of the statistical analysis is to identify these few important experimental factors.

If there are no restrictions on sample size, we can use replicated observations to provide an independent estimate of experimental error. This estimate could then be used to help us determine which effects are large enough to be considered important signals. However, as for many biotechnology problems, we were not able to collect replicate observations in this example. Thus, we relied on special graphical methods for the analysis of unreplicated fractional factorials.

Generally, a regression program is used to estimate the factor effects. These effects (or the coefficients) are then displayed in a Pareto chart and a normal plot. The Pareto chart shows the effects in order of their absolute magnitudes and helps separate the "vital few" from the "trivial many" effects." The normal plot is constructed so that the points associated with important effects fall away from a line drawn through the rest of the points. This method identifies the most important signals.

An active contrast plot is another useful method for identifying important effects. It is based on the effect sparsity assumption and provides estimated (posterior) probabilities that each effect is active (important). This probability is the chance that the effect is active given the sizes and spread observed among all of the estimated effects. Probabilities of over .50 are usually considered as indicative of important effects. Other graphical methods such as cube plots and interaction plots can help us understand the practical meaning of the effects.

Once a reduced model is selected, we fit the model and examine the ANOVA table. A t-value is calculated for each factor or interaction by dividing its coefficient by the estimated standard error. This is similar to a signal-to-noise-ratio so that important effects result in large t-values. p-values may be used to evaluate the significance of the t-values. Small p-values are generally associated with important effects. The R-squared value provides an overall measure of the fit of the model. Several diagnostic plots of predicted values and residual values can be used to identify potential problems with the fit of the reduced model.

The graphical and analytical methods described in this chapter provide a powerful means of interpreting the results from small, unreplicated fractional factorial experimental designs. These designs are especially chosen to be small and efficient. The use of proper methods for analysis and presentation of the results maximizes the return on the time and money spent on collecting experimental data.

References and Bibliography

Box, G. E. P. and N. R. Draper (1987). *Empirical Model-Building and Response Surfaces*. New York: Wiley.

Box, G. E. P., W. G. Hunter, and J. S. Hunter (1978). *Statistics for Experimenters: An Introduction to Design, Data Analysis, and Model Building*. New York: Wiley.

Box, G. E. P. and R. D. Meyer (1986). An analysis for unreplicated fractional factorials. *Technometrics*, **28**, 11-18.

Carroll, R. J. and D. Ruppert (1988). *Transformation and Weighting in Regression*. New York: Chapman and Hall.

Cleveland, W. S. (1985). *The Elements of Graphing Data*. Monterey, CA: Wadsworth.

Daniel, C. (1959). Use of half-normal plots in interpreting factorial two-level experiments. *Technometrics*, **1**, 311-342.

Daniel, C. (1976). *Applications of Statistics to Industrial Experimentation*. New York: Wiley.

Dixon, W. J. and F. J. Massey, Jr. (1983). *Introduction to Statistical Analysis*, 4th Edition. New York: McGraw-Hill.

Draper, N., and H. Smith (1981). *Applied Regression Analysis*, 2nd Edition. New York: Wiley.

Mallows, C. L. (editor) (1987). *Design, Data, and Analysis by Some Friends of Cuthbert Daniel*. New York: Wiley.

Nachtsteim, C. J. (1987). Tools for computer aided design of experiments. *Journal of Quality Technology*, **19**, 132-160.

Quality and Productivity Improvement Using the SCA Statistical System (1987). Scientific Computing Associates, P. O. Box 625, DeKalb, Illinois 60115.

SAS (1988). SAS Institute, Inc., SAS Circle, P.O. Box 8000, Cary, NC 27512-8000.

Snedecor, G. W., and W. G. Cochran (1980). *Statistical Methods*, 7th Edition. Iowa State.

SCA (1987). Scientific Computing Associates, P. O. Box 625, DeKalb, Illinois 60115.

SPlus (1988). Statistical Sciences, Inc., P. O. Box 85625, Seattle, WA 98145-1625.

Statgraphics (1987). STSC, Inc., 2115 East Jefferson Street, Rockville, Maryland 20852.

Weisberg, S. (1985). *Applied Linear Regression*, 2nd Edition New York: Wiley.

Chapter 5

SELECTING AN EXPERIMENTAL DESIGN

A well-designed experiment uses the fewest number of measurements to get the greatest amount information; that is, it is small and efficient. We have developed a set of these small, efficient experimental designs which are very powerful, widely applicable, and especially well suited for biotechnology research and development. These designs and their properties are described in detail in the Design Digest at the end of this book. In this chapter we compare the properties of different types of experimental designs and show how to select the best design for a particular experiment.

5.1   Types of experimental designs

By looking in the Design Digest, we see that there are many designs to choose from for any particular experiment. A complete listing is given in Table 5.1. For example, suppose that we wish to conduct an experiment with four factors. The Design Digest lists the following four factor designs: FF0416, FF0408, IF0412, MF0412, and MF0424.

In this notation, the first two letters of a design's name tell what type it is, the first two digits tell how many factors, and the last two digits give the sample size. Thus, these designs are, respectively, sixteen run and eight run two-

85

level fractional factorials (FF), a 12 run irregular fraction (IF), and twelve run and twenty-four run mixed two- and three-level fractional factorial (MF) designs. The different properties of these designs help us to decide which is best for a given experiment.

Table 5.1 Listing of Experimental Designs in the Design Digest

	Design Name				
Number of Factors	8 Run Designs	12 Run Designs	16 Run Designs	24 Run Designs	32 Run Designs
3	FF0308	MF0312	*	*	*
4	FF0408	IF0412 /or MF0412	FF0416	MF0424	*
5	FF0508	IF0512/or MF0512	FF0516	MF0524	FF0532
6	FF0608	IF0612	FF0616	IF0624	FF0632
7	FF0708	PB0712	FF0716	PB0724	FF0732
8	+	PB0812	FF0816	PB0824	#
9	+	PB0912	FF0916	PB0924	#
10	+	PB1012	FF1016	PB1024	#
11	+	PB1112	FF1116	PB1124	#

* This design is not in the Design Digest, consider replicating a smaller design.

This design is not in the Design Digest, consider using a 24 run design or contact a statistician for assistance

+ No design exists for this combination of factors and runs.

KEY: Design Name = aakkrr where
        aa=FF (two-level fractional factorial),
           MF (mixed two- and three-level fractional factorial),
           IF (irregular two-level fractional factorial),
           PB (Plackett-Burman)
        kk=number of factors, kk=3, 4, 5, ..., 11
        rr=number of runs, rr=8, 12, 16, 24, 32.

Two-level fractional factorial designs: Resolution V

Resolution V fractional factorial designs were discussed in Chapter 3. We call these designs "clear signal" designs because they allow the independent estimation of all main effects and all two-factor interactions. For a given number of factors, the two-level, Resolution V fractional factorial design is usually the largest design listed in the Design Digest. These designs are as follows: FF0308, FF0416, FF0516, FF0632.

Note that the three and four factor designs are full factorials whereas the five and six factor designs are each one-half fractions. These clear signal designs are especially useful when two-factor interactions are likely, there are no severe restrictions on sample size, and there are six or fewer experimental factors. There are no two-level Resolution V fractional factorials included in the Design Digest for seven or more factors because these designs are too large for most applications.

Two-level fractional factorial designs: Resolution IV

Although Resolution IV designs do not allow independent estimates of all the two-factor interactions, they do provide independent estimates of all of the main effects. Thus, a Resolution IV design makes a trade-off between sample size and information about two-factor interactions.

The following designs in the Design Digest belong to this group: FF0408, FF0616, FF0716, FF0732 and FF0816. There is no comparable design for five factors. The designs for nine or more factors are too large for consideration here.

The specific information about confounding of effects is given in the design's confounding pattern. For example, the eight run fractional factorial design, FF0408, has the following confounding pattern (see the Design Digest):

$$A*B + C*D$$
$$A*C + B*D$$
$$A*D + B*C$$

In this listing we assume that A is the first factor, B is the second factor, etc. (Note: for simplicity, all alias patterns are listed with "+'s" although in some cases a "-" is technically correct.)

The line $A*B + C*D$ means that the two-factor interaction $A*B$ cannot be estimated separately from that of $C*D$. What is really being estimated is the sum (or difference) of the two interactions. In this case, we say that $A*B$ is aliased with $C*D$.

If we are using a computer program to estimate the effects, we can enter either $A*B$ or $C*D$ into the model but not both. Since there are three lines to the confounding pattern, we can estimate all of the main effects plus three two-factor interactions. We can pick one name from each line for the three interactions; for example, A*B, A*C, and A*D. It doesn't matter which names we pick from each line.

Because of this confounding, we cannot tell from the data whether the estimated effect of A*B + C*D is due to A*B, C*D, or to some combination of the two. This is the information we give up in order to use a smaller sample size. Usually, we pick the interaction name which seems most plausible based on scientific knowledge of the problem. We assume that factors are listed from most to least important, so the two-factor interaction between the highest pair of factors is always listed first in each line of the confounding pattern.

If it becomes important to understand which of the two-factor interactions has caused the effect, there are several possible remedies. Box, Hunter and Hunter (1978) describe how to conduct a second experiment which can be added to the first to remove some of the confounding among of the interactions. These designs are called reflected designs and are constructed by reversing the high and low levels of one or more of the factors. The two experiments can then be analyzed together to get independent estimates of the some of the two-factor interactions. For additional approaches of this type see the SCA handbook *Quality and Productivity Improvement Using the SCA Statistical System (1987)*.

A second approach to deciphering confounded interactions is to run only one or two more points which can be selected by an optimal design program. For further information about this approach, see the following discussion on computer generated designs and also the E-Chip manual (1986). This method picks the fewest possible points needed for a specific interaction but will generally provide very little information other than what was specifically requested. In comparison, a reflected design requires more runs but also provides much more information.

If we are trying to further improve process performance, a second experiment can be designed with better factor settings which will also provide more information about interactions. In particular, the confounding pattern of the second experimental design can usually be arranged to separate the two-factor interactions in question. If one or more factors are no longer thought to be important, the second experiment will also provide added information about two-factor interactions. This method allows continued progress toward the optimum process settings and provides for a possible reduction in the number of experimental factors.

Two-level fractional factorial designs: Resolution III

The two-level, Resolution III fractional factorial designs are useful in the early stages of an investigation because they offer the opportunity to investigate many factors using only very few runs. In order to accomplish this, Resolution

III designs allow confounding not only among two-factor interactions but also among main effects and two-factor interactions. These designs are appropriate if we believe in advance that the two-factor interactions are not likely to be large in comparison to the main effects. The designs of this type in the Design Digest are FF0508, FF0608, FF0708, FF0916, FF1016, and FF1116.

A Resolution III design may allow the estimation of a few two-factor interactions. However, the two-factor interactions will be aliased with other two-factor interactions. For example, the 8 run, 5 factor design (FF0508) has the following confounding pattern (factor names are A - E):

$$A + D*E$$
$$B + C*E$$
$$C + B*E$$
$$D + A*E$$
$$E + B*C + A*D$$
$$A*B + C*D$$
$$A*C + B*D$$

Each of the five main effects is confounded with one or more two-factor interactions, for example, A is aliased with D*E. The remaining two-factor interactions are confounded only with other two-factor interactions and not with main effects, for example, A*B and C*D.

This design is not appropriate if there is an interaction between factors D and E because D*E is confounded with A. This design is suitable if A*B and A*C are the only large two-factor interactions. If we aren't sure which two-factor interactions may be important, then it is better to use a Resolution IV or V design, for example FF0516.

When using this (or any of the designs in the Design Digest), it is assumed that factors are listed from most to least important. With this in mind, FF0508 confounds A, the most important main effect, with D*E, the interaction between the two least important effects, and so on. If factors are listed in order of importance then it is reasonable to think that A*B is the most likely two-factor interaction followed by A*C, etc. Different confounding patterns can be obtained by rearranging the order in which the factors are listed.

Plackett-Burman designs

The Plackett-Burman designs (Plackett and Burman, 1946) included in the Design Digest allow the investigation of many factors using very few measurements. The 12 run Plackett-Burman designs included in the digest are Resolution III designs. They require fewer runs than a comparable fractional factorial design. For example, we can investigate 11 factors with a 12 run Plackett-Burman (PB1112) compared to the 16 run fractional factorial alternative

(FF1116). (Note: in some cases, the Plackett-Burman and two-level fractional factorial design are the same, for example, FF0708.) The twelve run Plackett-Burman designs are PB0712, PB0812, PB0912, PB1012, and PB1112.

Plackett-Burman designs are very useful for picking the one or two most important factors from a long list of candidate factors. At this early problem solving stage, we assume that important main effects will be much larger than two-factor interactions so we are willing to confound main effects with two-factor interactions.

In addition to the standard 12 run Plackett-Burman designs, the Design Digest includes several 24 run Plackett-Burman fold-over designs (Box and Wilson, 1951); namely, PB0724, PB0824, PB0924, PB1024, and PB1124. These are all Resolution IV designs and allow the estimation of all main effects and some two-factor interactions. They are useful when a larger sample size is possible and a design with higher power is required. (See the discussion on power in Section 5.2.)

Nonorthogonal designs: two-level irregular fractions

Irregular fractions of factorial designs provide an alternative to standard fractions which provide higher resolution with a smaller sample sizes. P. W. M. John (1969) proposed a class of these designs which are also discussed in Diamond (1981). The Design Digest includes the following designs by John: IF0412, IF0512, and IF0612. A similar design, IF0624, (developed by R. F. Liddle) is also included in the digest.

Irregular fractional factorials cannot be strictly classified by resolution in the same way as standard fractional factorials. Consequently we say that designs IF0412 and IF0624 are *nearly* Resolution V and IF0512 and IF0612 are *nearly* Resolution IV. This is because some effects may be slightly confounded and no longer completely independent. The classification of *nearly* Resolution IV or V is based on an analysis of the degree of partial confounding. Little is lost if a standard analysis is carried out and the slight partial confounding levels are ignored. (See Section 5 of the Design Digest.)

Significant savings in sample sizes can be achieved using these designs. For example, the twelve run design IF0412 is a 3/4 (=12/16) fraction of the sixteen run full factorial (FF0416). It is a compromise between the eight run (FF0408) and sixteen run fractional factorials. In comparison to FF0408, the four extra runs increase the resolution from III to V. Yet FF0412 saves four runs and still has nearly the same resolution as the 16 run design. Similarly, IF0512 is a 3/8 fraction and provides an intermediate design between FF0508 and FF0516.

Mixed two- and three-level fractional factorial designs

Mixed two-and three-level designs are similar to two-level fractional factorials. However, these designs allow one factor to have three levels (low, middle and high) instead of just two levels. They are particularly useful when one of the factors under consideration may have its best setting in the middle of the range of interest.

The designs of this type in the Design Digest are MF0312, MF0412, MF0512, MF0424 and MF0524 (Liddle and Haaland, 1988). The designs MF0312 and MF0424 are full factorials (3*2*2=12 and 3*2*2*2=24). The rest of these designs are fractions; for example MF0412 is a one-half fraction (12/24 = 1/2). Additional information on mixed two- and three-level fractional factorial designs (also called asymmetric fractional factorials) can be found in Addelman (1962), Margolin (1968, 1969), Daniel (1976), Taguchi and Wu (1980), Dey (1985), and Taguchi (1986, 1987).

The mixed-level factorial designs which are not full factorials are nonorthogonal designs and so cannot be strictly classified according to resolution. As in the case of irregular fractions, these designs are very nearly standard resolutions. For example, MF0512 is nearly Resolution IV when the three-level factor is quantitative (that is, there are no two-factor interactions between the three-level factor and other main effects). MF0412 is nearly Resolution V when there are no two-factor interactions between the three-level factor and other effects.

These designs allow the estimation of a quadratic effect in the statistical model to allow for curvature in the three-level term. This allows the middle setting to have a higher (or lower) response value than either the low or high setting. The curvature term is listed as A*A (A-squared) in the Design Digest. We discuss the analysis of these designs more fully in the next chapter. (See also the discussion in the Sections 5-7 of the Design Digest.)

Computer generated designs

There are many software packages which generate experimental designs. Nachtsheim (1987) provides a comprehensive review. Many of these software packages primarily generate standard two-level fractional factorials. However, some packages, E-Chip (1986) for example, generate "optimal" designs. These designs are optimal according to certain statistical design principles. (See, for example, Khuri and Cornell, 1987, for a discussion of optimal design criteria.)

One popular optimal design criterion is D-efficiency. A D-optimal design maximizes information about the parameters for a specified model and a given number of observations. In selecting a D-optimal design, the user specifies the model which he/she believes to be true. The model includes main effects and any interactions of interest. For a given sample size, the computer program finds the best combinations of experimental conditions to run. Most of the designs listed in the Design Digest are D-optimal for their sample size and fully specified model (i.e., all main effects and estimable two-factor interactions).

One way in which D-optimal design programs are often used is to include in the model the main effects and only the two-factor interactions of particular interest. Then the smallest design is found which estimates these effects. There are two drawbacks to this approach. First, if there are unexpected two-factor interactions, the design will no longer provide good estimates of the effects. Secondly, if one of the observations cannot be made for some reason, the design loses much of its efficiency.

Thus, the smaller D-optimal designs may not be robust, and we do not advocate their use for models containing only a few interactions. The standard designs included in the Design Digest (which are usually D-optimal for the maximum number of two-factor interactions) provide a more conservative approach which we believe is robust to model misspecification and occasional missing data points. These designs also provide better coverage of the experimental space.

A recent development in computer software is the use of programs which assist the user in evaluating available designs. See, for example, Haaland, et al (1985, 1986, 1988) for a description of an expert system called Dexter which guides a user in selecting the best design. This program examines the problem as defined by the user and then tries to pick the design which best matches the user's needs. Such programs are useful in evaluating alternative designs and can make it easier for the user to choose from among the many designs which may be available.

Summary

The Design Digest is a collection of small, efficient designs which are useful for biotechnology product research and development. A complete listing of design names by sample size and number of factors is given in Table 5.1.

These designs cover a wide range of resolution, sample size, number of factors, and confounding patterns. Some designs are appropriate for a first experiment which must find the few important factors from among a long list of candidate factors. Other designs are appropriate for examining a few factors and identifying all of their important two-factor interactions. Each of these designs may be effectively used in problem solving to identify important factors and to move toward the optimal settings of the important factors.

5.2   Power and sample size

The power of a design is its ability to correctly identify important factors; that is, a design with low power is more likely to overlook one or more important factors than a design with higher power. However, in order to increase the power of an experiment, we need to collect more data. Thus, we must trade off the need for high power against the cost of collecting more data.

The power of a design determines how large a signal we can expect to detect against a certain level of background noise. The larger a signal is, the more likely we are to find it. In contrast, the higher the background noise the more difficult it is to detect smaller signals. The size of the signal is determined by the process. However, by collecting more data we can decrease the noise level and so increase the power. In fact, the noise level (standard error) decreases roughly as a factor of the square root of the sample size.

Thus, increasing the sample size is the standard way for conducting a more powerful experiment. One way to increase sample size is by using a larger design; or example, a sixteen run design versus a twelve run design. Another way is by replication, which usually means repeating the entire experiment one or more times. For example, an experiment planned for one microtiter plate can be replicated on a second plate or the experiment could be repeated on a second day.

Type I, Type II, and Type III errors

There are two commonly considered errors which may occur when we evaluate the significance of a particular experimental factor. A Type I error occurs if an unimportant factor is called a signal (false positive). A Type II error occurs if an important factor is not identified as a signal (false negative). The risk of making a Type I error is called the α level. The risk of making a Type II error is called the β level. The power is one minus the probability of a Type II error (1-β). Hence the power is the chance of correctly identifying an important factor as a signal.

We speak of the power of a design in connection with the significance of a factor (or interaction term). In this sense, the power is the sensitivity of the design for finding important factors. Analogously, the α or significance level, is the specificity of the design for finding only important factors. For example, a design may have power 80% to detect a particular effect at a 5% significance level. In this case, there is a 5% chance of making a Type I error and a 20% (= 100%-80%) chance of making a Type II error. A significance level of 5% corresponds to a 1 in 20 chance of a false positive. A power of 80% corresponds to a false negative rate of 20% which represents a 1 in 5 chance of missing an important factor.

A Type III error is more difficult to quantify than the first two types. It is defined as "answering the wrong question". For example, a Type III error arises if the most important variable affecting process performance isn't included in the experimental design. Another Type III error occurs if, for example, process yield is greatly improved but the resulting loss of purity prevents successful extraction of the target ingredients. Unfortunately, a Type III error cannot be prevented by increasing the sample size or choosing a better design. The best prevention for Type III errors is good communication and clear thinking.

Determining sample size

Table 5.2 shows the (approximate) sizes of effects which can be detected at different significance and power levels for fractional factorial type designs. The table entries correspond to the standard design sizes used in the design digest. The entries in the table are effect magnitudes divided by the size of the experimental error (standard deviation of an individual observation). Thus, the value 1.4 means that a 16 run two-level fractional factorial can detect an effect 1.4 times as large as the experimental error with a power of 80% and significance level of 10%. The table is adapted from Nelson (1985).

We typically prefer 10% significance and 80% power in selecting a design. The 80% power provides a reasonable assurance of detecting an important effect which is intermediate in size. The 10% significance level takes into account the high cost of missing an important factor compared to carrying along an unimportant factor into the next experiment. At the expense of using larger samples, it is possible to use the 5% significance level or the 90% power level to provide higher protection against false positives and false negatives, respectively.

The entries in Table 5.2 are expressed in units of experimental error (of the individual observations) because we typically know only the relative sizes, rather than the actual values, of the effects and the experimental error. For example, we may want to identify any effect which is twice as large as the experimental error. This is a reasonable assumption for an important factor in a screening experiment. However, in the later stage of an investigation, we may need more precise knowledge and so wish to identify an effect which is only as large as the standard error.

In order to determine the proper sample size using Table 5.2, we choose a significance level and power, for example 10% and 80%, respectively. Then we see that an eight run, two-level fractional factorial will detect an effect twice as big as the experimental error. However, a 32 run design is required to detect an effect smaller than the experimental error. As the sample size increases, we have the power to find smaller and smaller effects.

In contrast, for a fixed sample size, say 16 runs, and significance 10%, we are more certain of detecting an effect which is 1.6 times the experimental error (90%) than of detecting an effect 1.4 times the experimental error (80%).

For a sixteen run design with fixed power, say 80%, as we decrease the significance level, we lose the ability to detect smaller effects. For example at 10% significance the minimum detectable effect is 1.4 whereas it is 1.6 for 5%. By decreasing the significance, we are less willing to allow a false positive so an effect must be bigger before we call it important.

Table 5.2 also provides sample size information for designs with a factor at three levels. We have only 12 and 24 run designs of this type. The table en-

Table 5.2 Power and Sample Size for Fractional Factorial Designs

Minimum Detectable Effect*
(Effect / Experimental Error)

	Sample Size				
	8	12	16	24	32
	significance = 10%			power= 80%	
two-level	2.0	1.6	1.4	1.1	.9
three-level		2.5		1.5	
	significance = 10%			power = 90%	
two-level	2.5	2.0	1.6	1.4	1.1
three-level		3.0		1.8	
	significance = 5%			power = 80%	
two-level	2.5	2.0	1.6	1.2	1.0
three-level		3.0		1.8	
	significance = 5%			power = 90%	
two-level	3.0	2.3	1.8	1.4	1.2
three-level		>3.0		2.0	

*   Values in this table are approximate. This table was adapted by permission of the American Society for Quality Control from Nelson (1985).

tries show that these designs are less powerful than the comparable two-level designs. This is because we have less information at each level of the three-level design. However, the mixed two- and three-level design has the advantage of providing information about the interior of the experimental region which may be very important in some cases.

Conclusion

We prefer not to use designs with power less than 80% because they are likely to waste time and money. For example, an eight run design has a very small chance of detecting a signal which is only as large as the experimental error of an individual observation. This doesn't mean that small experiments are not useful; just that they should be used wisely. In particular, an improvement in process performance which is one-and-one-half to two times as large as the experimental error is a realistic goal for a screening experiment. It is prudent to carefully assess costs and to use the largest design (highest power) which is consistent with experimental objectives and resources.

5.3   Replication

Replication increases the power of an experiment without using additional design points by making repeated observations at existing design points. Replication allows for an independent estimate of the experimental error and is often less expensive than using a larger design (i.e., more design points). For example, it may be easier and less expensive to repeat a sixteen run design on a second plate rather than to prepare thirty-two different samples.

Independent replicates

Care must be taken to ensure that the replication involves independent data values -- not just repeat readings -- at each design point. For example, running an analysis on the same sample more than once would not be useful for increasing the power unless the measurement process itself is highly variable. However, an experiment repeated on two separate days provides true replicates and greatly increases the power because the sample size is doubled.

An important advantage of replication is that it increases the reliability of information obtained at each design point. In contrast, a larger design (with more design points) provides more information about the experimental region, but it doesn't increase information at any single design point. Thus, replication is preferable to a larger design if sample-to-sample variability is high. Otherwise, a larger design is preferred because it more thoroughly explores the experimental region.

Replicate observations have an special benefit for the statistical analysis; namely, they allow an independent estimate of experimental error. This error estimate is based on the standard deviation among replicates. Such an estimate increases the validity of formal statistical tests and lessens reliance on informal graphical means of analysis. Replication also avoids errors because nonsignificant terms do not need to be dropped from the model in order to provide an estimate of the experimental error.

Averaging multiple observations

Multiple observations which take, for example, the form of duplicate or triplicate wells on the same plate, several mice which are treated at the same time and housed together under similar conditions, aliquot samples, etc., are usually averaged together before analysis. Averaging the values reduces the variability and so increases the power.

In distinguishing between independent replicates and multiple observations (which should be averaged), there is usually a gradient. For example, experiments run on two different microtiter plates which were set up and run on different days would be truly independent replicates. Two plates run in parallel under the same conditions are less independent. Finally, multiple wells on the same plate produce measurements which are not usually independent. A useful rule of thumb is to average over wells but to treat different plates as independent replicates.

Independent replicates may be treated as direct additions to the sample size when using Table 5.2. They are also treated as separate data values in the statistical analysis. However, replicates which are not independent, as in the case of duplicate wells, should be averaged before the analysis. In this case, increased power results not from a larger sample but from reduced variability.

The primary drawback to repeated observations (which have been collected under similar conditions so that they are not completely independent) is that the main component of experimental error (usually from setup-to-setup) is not reduced. For example, plate-to-plate variability is usually greater than well-to-well. However, this method is the least expensive way to increase the power of an experiment. It will be most advantageous when the repeated observations represent a fairly large fraction of the total experimental error.

5.4   Center points

A center point is an observation or run for which all factors have an intermediate value; that is, in the middle between the "high" and "low" values. Center points increase power by providing a better estimate of the experimental error.

Center points also indicate whether the best conditions are in the interior of the current experimental region which suggests rapid progress toward optimization.

Each center point is run under exactly the same conditions. Differences in the measured values of the response among the center points should be due solely to experimental error. Thus, the standard deviation of the center points provides an independent estimate of the experimental error.

The Design Digest recommends a number of center points for each design. For example, the listing for a five factor, sixteen run design (FF0516) includes 3 center points. Center points are not recommended for Resolution III designs or for the 12 run designs since we assume they are being used because of sample size restrictions. Section 6 of the Design Digest provides a more detailed discussion of center points. Section 7 discusses the use of center points for detecting curvature (for example, if the best response occurs in the interior of the experimental region). See also the example in Chapter 7.

5.5 Randomization and blocking

Systematic bias arises from uncontrolled changes in experimental conditions. The presence of systematic bias may lead to confusing or even incorrect interpretation of the data. Two very effective means of preventing systematic bias are randomization and blocking.

Sources of systematic bias

Statistical methods were developed to deal with variation or noise in data. However the statistical analysis assumes that the variation or error in the data follows a random pattern. In particular, we assume that the average error is zero and that the pattern of errors follows a normal probability distribution. A systematic bias violates these assumptions.

Systematic bias arises when conditions change during the course of an experiment in such a way that the changes are confounded with the estimates of the factor effects or the experimental error. Potential sources of systematic bias include

- data collected by more than one lab technician,
- responses measured on more than one day,
- more than one batch of raw material used,
- raw materials degraded over the course of an experiment,
- samples degraded during the measurement process,
- equipment changed during the experiment, and
- samples not stored under uniform conditions.

These or other systematic biases may make the results of an experiment inconclusive or even misleading.

For example, suppose two technicians, one of whom generally gets higher readings, are making measurements. Now suppose that one of these technicians collected all of the data associated with high setting of one factor and the other technician collected all of the readings at the low level. Then the difference in average readings between the two technicians is confounded with the effect of the experimental factor.

To control possible bias

- all materials and methods should be homogeneous and
- all data should be collected over a short time period.

Randomization and blocking provide further tools to minimize systematic bias.

Blocking

If, for example, more than one batch of raw materials or more than one technician must be used for an experiment, this source of possible bias can be controlled by modifying the experimental design. One useful method is called blocking.

Blocking associates the experimental condition which changes with a term in the model, usually a two- or three-factor interaction. Most computer programs for generating designs provide this facility (see Nachtsheim, 1987). If a design package does not allow blocking, the blocking factor can be treated as an additional experimental factor. (This assumes that the blocks will not interact with the other factors.) More complicated blocking procedures can be obtained from Box, Hunter and Hunter (1978). The SCA handbook (1987) also provides additional details.

To run an experiment in two blocks, one of the two blocks is determined by the high settings of the blocking factor and the other block is determined by the low settings. For example, the five factor, sixteen run fractional factorial (FF0516) can be run in two blocks using the lowest two-factor interaction, D*E. Multiply the columns for D and E together and let the runs with minuses be the first block and the runs with pluses be the second block. Thus, runs 1, 2, 7, 8, 11, 12, 13, and 14 are in the first block. Runs 3, 4, 5, 6, 9, 10, 15, and 16 are in the second block. If center points are included, two should be run in each block.

For example, suppose all of the samples for a sixteen run experiment cannot be prepared on one day. In order to control for day-to-day changes in experimental conditions, we can block on day. Then half of the experiment is prepared each day. For design FF0516, as described above, we can use the two-factor interaction D*E as the blocking factor. The runs from the first block are prepared on the first day and those in the second block are prepared on the second day. However, the results from both blocks are combined for the analysis.

The blocking assures that preparation differences from day-to-day do not affect the estimates of the rest of the effects. The same method can be applied for two batches of a raw material, two technicians, two instruments, etc. An example of blocking used to control for day-to-day effects is presented in Chapter 9.

The advantage of blocking is that the blocking factor is independent of all the other factors. Any difference between the two blocks won't affect the estimates of the other model terms. An estimate of the effect due to blocks can be found by putting the blocking term (D*E in the above example) into the statistical model. Thus, blocking effectively prevents systematic bias when changes in experimental conditions cannot be avoided.

Randomization

Randomization is a simple and powerful tool for preventing systematic biases due to unanticipated changes in experimental conditions. An experiment is randomized by setting up the experimental conditions, assigning samples to storage conditions, making measurements, etc. in a randomized order.

Randomization prevents a systematic bias because it is very unlikely that unanticipated changes in experimental conditions will correspond exactly to the random pattern in which the experiment is run. We randomize even when all known sources of experimental error have been identified and controlled because it is cheap protection against some chance occurrence ruining the results of an entire experiment.

For example, we prefer to randomly assign runs to different lab instruments, technicians, days, etc., even if we strongly believe that there should be no differences in results. If randomization is extremely difficult or impossible, extra care should be taken to run the entire experiment under uniform conditions.

The actual procedure for randomization is quite simple. Most commercially available design packages provide this facility. If a computer isn't available, the manual procedure is as follows: use a random number table to select a sequence of integers between 1 and n where n is the sample size, then set up and run the experiment in that order. For example, if the first random number is 7, then the seventh design point is run first (or, alternatively, the first design point is run seventh) and so on.

Summary

Experimental results are said to be *reproducible* if similar results can be obtained in subsequent experiments. Because of experimental error, this does not mean that the same measurements will be obtained, only that we can reach the same conclusions. Preventing systematic bias is a key element in obtaining reproducible results because it helps guard against "unexplainable" patterns in the experimental results.

In addition to randomization and blocking, reproducible results depend on practical considerations such as

- not overlooking important experimental factors and
- having good communication among project team members

If there is a problem with reproducibility which cannot be solved with the methods of this section, a statistician may help you design a special study to identify your sources of variability (See for example the discussion on estimating variance components in Snedecor and Cochran, 1980).

5.6   Example: Cell culture system

To further illustrate how to choose an experimental design, we consider the empirical optimization of a cell culture system which is used to produce monoclonal antibodies. Recall that in Chapter 2 we discussed the use of animals to produce monoclonal antibodies. Cell culture systems provide an alternative method which allows the in vitro production of monoclonal antibodies. In cell culture systems, the antibody secreting cells are grown in culture and the monoclonal antibodies are harvested from the circulating medium in the system.

As in the use of animals to produce ascites, cell culture systems are complex, biological systems which require a systematic problem solving approach to find their optimal production settings. The purpose of this investigation is to establish whether or not variation of certain of the operating parameters affects system performance. We discuss several possible experimental designs for this problem and the circumstances in which they would be appropriate.

Problem description

A schematic of a cell culture system is shown in Figure 5.1. Each cell culture unit is a self-contained, complete entity consisting of a culture unit, a flow meter and in-line filter. For the experiments in question, one peristaltic pump can accommodate four culture systems at once. All units must pass an integrity test and are flushed with an ethanol solution prior to setup. During the course of an experiment, each system is housed in a 37°C incubator in a controlled atmosphere.

Each unit is to be inoculated with clone cells in medium with or without serum as specified by the experimental design. The recirculating culture medium in the reservoir of each system is identical (except in volume) for each system

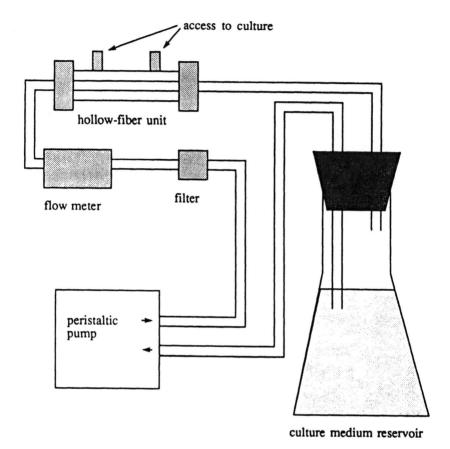

Figure 5.1 This schematic for a cell culture system describes a system for the in vitro production of monoclonal antibodies. Each system is self contained. Monoclonal antibodies are harvested from the extracapillary space. The culture medium is circulated by means of a peristaltic pump at a controlled rate.

and is replaced once a week. Harvests are collected from the extracapillary space.

The cell culture units will be randomly assigned to the experimental treatment conditions. The units will be operated for a period of five weeks under the conditions laid out by the experimental design. Viable cell concentrations are determined using an ethidium bromide-acridine orange staining method. Specific antibody concentrations are determined using a liquid phase radioimmunoassay specific for the monoclonal antibody. Antibody yields will be calculated from specific antibody concentrations and total harvest volume.

Experimental factors

The following eleven operational parameters (listed in order of suspected importance) may affect system performance and therefore are potential factors for inclusion in this study:

- harvest frequency,
- harvest volume,
- nutrient media reservoir size,
- the presence of fetal bovine serum in the cell inoculum,
- nominal molecular weight cut-off of the membrane,
- flow rate,
- membrane type,
- hollow fiber unit packing,
- membrane surface to extracapillary volume ratio,
- harvest recycling, and
- fetal bovine serum in the recirculating medium.

We think that the first five factors are critical to process performance. The second four factors are probably important to process performance. The last two factors may be important.

Although we are not sure about all of the two-factor interactions, interactions among the first five factors seem likely. At least some of these interactions could be as important as the effects of the factors themselves. Interactions among other factors are uncertain but possible.

Selecting a screening experiment design

The first experiment is a screening experiment to identify important factors affecting productivity and to suggest improved settings for the operational parameters. A maximum of sixteen cell culture units are available for use.

There are two eleven factor designs in the Design Digest; namely, FF1116 and PB1112. The Plackett-Burman design provides the smallest sample size. However, the fractional factorial design allows the estimation of five two-factor interactions. Because the Plackett-Burman design is a smaller design, it is less powerful. Table 5.2 shows that with significance 10% and power =80%, a twelve run design can detect a difference of 1.6 standard deviations whereas a sixteen run design can detect a difference of 1.4 standard deviations. Which design we choose depends on the expense of running twelve extra runs versus the importance of being able to detect a smaller effect. The larger design, however, also provides additional information about two-factor interactions.

Since this is the first experiment, we may not be very close to the optimal settings. Since the experiment is costly and time consuming, we wish to use a small design until we are sure that we are fairly close to good settings for the process variables. Therefore, the twelve run Plackett-Burman design seems to provide the most economical approach. After running this design, we may be able to remove from consideration the last two (thought to be least important) factors. We should be able to choose better settings for all of the remaining factors.

Designing a second experiment - Scenario I

Suppose that after the first experiment there are eight process variables which may still have an important effect on antibody yield. We also feel that process performance is not close to optimal levels yet so interactions are likely to be smaller than main effects. In this case, it makes sense to conduct a second Plackett-Burman experiment (PB0812) in the region of the best settings found in the first experiment.

Designing a second experiment - Scenario II

Suppose that the results of the first experiment clearly show that only the first five factors are important. If the process yields for the best runs are close to target levels, the next design should provide estimates of the two-factor interactions among the first five factors. Table 5.1 shows that FF0516 is the two-level, clear signal design for five factors in 16 runs. Because it is only a Resolution IV design, the irregular fraction IF0512 provides a somewhat less desirable alternative.

Eight run alternative designs

In the first experiment, we assumed that 16 cell culture units were available. However, suppose that due to cost and limited space, approval could only be obtained to use eight units. Possible alternative approaches to the first experiment are as follows:

- Select the 7 most important factors and use design FF0708. This is a Resolution III design which would allow us to screen the greatest number of factors.
- Select the five most important factors and used FF0508, which is also a Resolution III design. This design would allow the estimation of two two-factor interactions (A*B and A*C).
- Select the four most important factors and use design FF0408. (For example, suppose that we had prior information which suggested the best setting for the nominal molecular weight cut-off of the membrane.) This

Table 5.3 Experimental Design Worksheet for Cell Culture Example

run*	Harvest Frequency hfreq	Media Harvest Volume (ml) hvol	Reservoir Volume (ml) medvol	Fetal Bovine Serum In Cell Inoculum (ml) fbsinoc	Yield† (mg) yield
1	weekly	0.5	100	yes	0.38
2	weekly	0.5	300	no	0.28
3	weekly	2.0	100	no	0.98
4	weekly	2.0	300	yes	1.06
5	daily	0.5	100	no	0.45
6	daily	0.5	300	yes	1.13
7	daily	2.0	100	yes	0.84
8	daily	2.0	300	no	2.70

* Runs were conducted in randomized order to guard against systematic bias.

† Due to contamination in unit 5 during the 5th week, the value from the fourth week alone (.45) was used for this analysis instead of the average from the fourth and fifth weeks.

design is a Resolution IV design and allows better protection against two-factor interactions.

• Keep all eleven factors as identified earlier, and run FF1116 in two blocks. (See Section 5.5.) This is like running two 8 run experiments and combining the results for analysis.

Which of the above alternatives is best depends on the circumstances. There is no one right answer. The actual experiment which motivated this example, used design FF0408 with the first four factors. The experimental design worksheet and results are shown in Table 5.3. The results of the analysis (which is left to the reader) showed that the best conditions were daily harvest frequency, 2.0 ml harvest volume, 300 ml media reservoir volume, and no fetal bovine serum in the cell inoculum.

Summary

In this example, we discussed some practical considerations in selecting an experimental design. We chose different designs depending on the assumptions we made about the process being studied. The designs we considered ranged from an 11 factor, 12 run Plackett-Burman design (PB1112) to a 4 factor, 8 run fractional factorial (FF0408). The range of alternatives illustrates the types of trade-offs we make when choosing an experimental design; namely, number of factors versus sample size and power versus resolution versus confounding patterns.

In reviewing design alternatives, we see that there is no hard and fast rule which provides the best design for every circumstance. This is why we provide a broad selection of small, efficient designs in the Design Digest. Based on scientific knowledge of the process being studied, on cost and time constraints, and on the information presented in the Design Digest, a researcher can select a suitable experimental design.

5.7   Review of selection procedure

The choice of the best design depends on the experimental objectives, the number of factors, the likelihood of two-factor interactions, and availability of resources. Consider the following rules of thumb for selecting the best design:

- Resolution III designs are useful for screening experiments which are intended to pick a few important factors from a long list of candidate factors.
- Resolution IV designs are useful for screening experiments in which some two-factor interactions may be important, there are limitations on sample size, and there aren't too many experimental factors.
- Resolution V designs allow independent estimates of all main effects and two-factor interactions and always provide a safe alternative when choosing a design. However, they may not be practical until the important factors have been identified by screening experiments.
- Use the highest resolution design which is affordable in order to protect against the possibility that unexpected two-factor interactions may lead to ambiguous results.
- The sample size chosen should be large enough to have a reasonable power level for the expected size of effects of the important factors.
- Use replication to reduce noise levels and increase power.
- Use center-points to determine whether the optimum system performance occurs near the center of the experimental region and to provide an independent estimate of the experimental error.
- Run the experiment in blocks, if necessary, to prevent biases due to expected changes in experimental conditions.
- Randomize the run order to protect against unexpected changes in experimental conditions.

Consult Box, Hunter and Hunter (1978), Diamond (1981), or Box and Draper (1987) for example, for designs not in the Design Digest.

Summary of selection procedure

Figure 5.2 shows a schematic of the procedure which we use to select an experimental design. The following rules provide a more detailed explanation of this strategy:

Step 1. Problem definition

A. Identify the experimental factors and list them in order of suspected importance.

B. Identify any factors which you want to investigate at three levels instead of two.

C. Assess the possibility of two-factor interactions. Identify known two-factor interactions.

D. Evaluate resources for sample size and determine the desired power. Use as a sample size guideline the number of observations that can be completed in a fairly short period of time with homogeneous materials.

Step 2. Consult the Design Digest

A. Identify the Resolution V design(s) available. Use a Resolution V design if possible.

B. If necessary, identify the Resolution IV design(s). Check the confounding pattern(s) to see if it is consistent with plans to estimate two-factor interactions.

C. If necessary, identify the Resolution III design(s). Review the likelihood of important two-factor interactions before using a Resolution III design.

Step 3. Evaluate the possible designs

A. Which of the designs are large enough to provide adequate power?

B. Should any new factors be added to the experiment? For example, FF0416 and FF0516 are both Resolution V so it may be more efficient to use five factors instead of four.

C. Should any factors be deleted from the experiment? For example, FF0916 is a Resolution III design whereas FF0816 is a Resolution IV design. Therefore, if the ninth factor is really marginal, the eight factor design may be preferred.

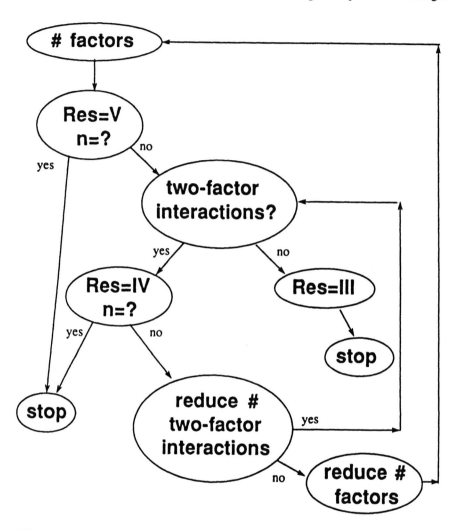

Figure 5.2 The process of selecting an experimental design involves evaluating the characteristics of candidate designs against the available experimental resources. The decision node indicated by "n=?" means "Do we have the resources to do an experiment of this size?" If so, use the Resolution V design. Otherwise, we must consider the likelihood of important two-factor interactions, etc. The flow chart is intended to direct experimenters to Resolution V designs if they have the resources. Resolution III designs are used if two-factor interactions are not important. Resolution IV designs provide a compromise which allows the estimation of some but not all two-factor interactions.

Step 4. Choose the best design

A. Make trade-offs among design properties as necessary to select the best design.

B. Don't use all of your resources in one experiment.

C. Consider whether there is a risk in dropping factors from the design or in ignoring the possibility of two-factor interactions.

Step 5. Consider additional design characteristics

A. Use blocking if there are any conditions which are expected to change during the course of the experiment.

B. Randomize run order to guard against unexpected changes.

C. Include center points if possible.

5.8   Conclusion

Selecting an experimental design is an important step in using statistical problem solving methods. The Design Digest, along with information about the experimental conditions, provides a powerful resource for choosing good designs. The designs in the Design Digest were included for their power and robustness and for their small size and efficiency. Additional improvements can be made in the reliability of experimental results by using randomization and blocking to guard against systematic biases.

References and Bibliography

Addelman, S. (1961). Irregular fractions of the 2^n factorial experiments. *Technometrics*, **3**, 479-496.

Addelman, S. (1962). Orthogonal main-effect plans for asymmetrical factorial experiments. *Technometrics*, **4**, 21-46.

Barker, T. B. (1985). *Quality by Experimental Design*. New York: Marcel Dekker and Milwaukee: ASQC Quality Press.

Box, G. E. P. and N. R. Draper (1987). *Empirical Model-Building and Response Surfaces*. New York: Wiley.

Box, G. E. P., and J. S. Hunter (1957). Multi-factor designs for exploring response surfaces. *Annals of Mathematical Statistics*, **28**, 195-241.

Box, G. E. P., and J. S. Hunter (1961a). The 2^{k-p} fractional factorial designs, Part I. *Technometrics*, **3**, 311-352.

Box, G. E. P., and J. S. Hunter (1961b). The 2^{k-p} fractional factorial designs, Part II. *Technometrics*, **3**, 449-458.

Box, G. E. P., W. G. Hunter, and J. S. Hunter (1978). *Statistics for Experimenters*. New York: Wiley.

Box, G. E. P. and K. B. Wilson (1951). On the experimental attainment of optimum conditions. *J. Royal Statistical Society, Series B*, **13**, 1-45.

Cochran, W. G. and G. M. Cox (1957). *Experimental Designs*, 2nd ed. New York: Wiley.

Daniel, C. (1976). *Applications of Statistics to Industrial Experimentation*. New York: Wiley.

Davies, O. L. (1956). *Design and Analysis of Industrial Experiments*, 2nd ed. New York: Hafner Publishing Company.

Dehnad, K. (1989). *Quality Control, Robust Design, and the Taguchi Method*. Pacific Grove, CA: Wadsworth & Brooks.

Deming, S. N. and S. L. Morgan (1987). *Experimental design: a chemometric approach*. New York: Elsevier.

Dey, A. (1985). *Orthogonal Fractional Factorial Designs*. New York: Wiley.

Diamond, W. J. (1981). *Practical Experimental Designs*. Belmont, CA: Lifetime Learning Publications.

E-Chip (1986). Expert in a Chip, Inc., RD1 Box 384Q, Hockessin, DE 19707.

Finney, D. J. (1945). The fractional replication of factorial arrangements, *Annals of Eugenics*, **12**, 291-301.

Ghosh S. (1987). Non-orthogonal designs for measuring dispersion effects in sequential factor screening experiments using search linear models. *Communications in Statistics - Theory and Methods*, **10**, 2839-2850.

Haaland, P. D., D. Yen, and R. F. Liddle (1985). An expert system for the design of screening experiments. *Proceedings of the Statistical Computing Section of the American Statistical Association*, 223-228.

Haaland, P. D., D. Yen, and R. F. Liddle (1986). An expert system for experimental design. *Proceedings of the Statistical Computing Section of the American Statistical Association*, 78-87.

Haaland, P. D., R. F. Liddle, J. C. Lusth, and D. S. Wilson, (1988). Dexter: An expert system for evaluating experimental design alternatives. *Proceed-

ings of the Statistical Computing Section of the American Statistical Association.

Hahn, G. J. and S. S. Shapiro (1966). A catalog and computer program for the design and analysis of orthogonal symmetric and asymmetric fractional factorial experiments. Report No. 66-C-165, General Electric Research and Development Center, Schenectady, New York.

Hahn, G. J. (1977). Some things engineers should know about experimental design. *Journal of Quality Technology*, 9, 13-20.

Hahn, G. J. (1982). Design of experiments: An annotated bibliography. In *Encyclopedia of Statistical Sciences*, Vol. 2, ed. S. Kotz and N. L. Johnson. New York: Wiley.

Hahn, G. J. (1984). Experimental design in a complex world. *Technometrics*, 26, 19-31.

Hendrix, C. D. (1979). What every technologist should know about experimental design. *Chemtech*, 9, 167-174.

Hicks, C. R. (1983). *Fundamental Concepts in the Design of Experiments*, 3rd ed. New York: Wiley.

John, J. A. (1987). *Cyclic Designs*. New York: Chapman and Hall.

John, P. W. M. (1969). Some non-orthogonal fractions of 2^n designs. *J. Royal Statistical Society, Series B*, 31, 270-275.

Kempthorne, O. (1979). *The Design and Analysis of Experiments*. New York: Robert E. Krieger Publishing Co.

Khuri, A. I., and J. A. Cornell (1987). *Response Surfaces: Designs and Analyses*. New York: Marcel Dekker, Inc. and ASQC Quality Press.

Liddle, R. F. and P. D. Haaland (1988). Efficient nonstandard screening designs. Presented at the Annual Meetings of the American Statistical Association, New Orleans, LA.

MacLean, R. A. and V. L. Anderson (1984). *Applied Factorial and Fractional Factorial Designs*. New York: Marcel Dekker.

Margolin, B. H. (1968). Orthogonal main effect 2^n*3^m designs and two factor interaction aliasing. *Technometrics*, 10, 559-573.

Margolin, B. H. (1969). Orthogonal main effect plans permitting estimation of all two factor interactions for the 2^n*3^m series of designs. *Technometrics*, 11, 747-762.

Mitchell, T. J. (1974a). An algorithm for the construction of D-optimal experimental designs. *Technometrics*, 16, 203-210.

Mitchell, T. J. (1974b). Computer construction of D-optimal first order designs. *Technometrics*, **16**, 211-220.

Montgomery, D. C. (1984). *Design and Analysis of Experiments*, 2nd ed. New York: Wiley.

Nelson, L. S. (1985). Sample size tables for analysis of variance. *Journal of Quality Technology*, **17**, 167-169.

Nachtsheim, C. J. (1987). Tools for computer aided design of experiments. *Journal of Quality Technology*, **19**, 132-160.

Patel, M. S. (editor) (1987). Experiments in factor screening. A special issue on group screening of *Communications in Statistics - Theory and Methods*, **10**.

Plackett, R. L. and J. P. Burman (1946). The design of optimum multifactorial experiments. *Biometrika*, **33**, 305-325.

Quality and Productivity Improvement Using the SCA Statistical System (1987). Scientific Computing Associates, P. O. Box 625, DeKalb, Illinois 60115.

SCA (1987). Scientific Computing Associates, P. O. Box 625, DeKalb, Illinois 60115.

Snedecor, G. W. and W. G. Cochran (1980). *Statistical Methods*, 7th Edition, Ames: Iowa State University Press.

Snee, R. D., L. B. Hare and J. R. Trout (1985) *Experiments in Industry: Design, Analysis, and Interpretation of Results*. Milwaukee: ASQC Quality Press.

Srivastava, J. N. (1975). Designs for searching nonnegligible effects. *A Survey of Statistical Designs and Linear Models*, J. Srivastava, editor. Amsterdam: North-Holland, 507-519.

Steinberg, D. M. and W. G. Hunter (1984). Experimental design: Review and comment (with discussion). *Technometrics*, **26**, 71-130.

Taguchi, G. (1986). *Introduction to Quality Engineering : Designing Quality into Products and Processes*. Tokyo: Asian Productivity Organization.

Taguchi, G. (1987). *System of Experimental Design*. Vol. 1 and 2. New York: Unipub.

Taguchi, G. and Y. Wu (1980). *Introduction to Off-line Quality Control Systems*. Tokyo: Central Japan Quality Control Association.

Welch, W. J. (1982). Branch-and-bound search for experimental designs based on D optimality and other criteria. *Technometrics*, **24**, 41-48.

Chapter 6

TRANSFORMATIONS

In this chapter, we discuss transforming the response variable to obtain a better fitting model. The statistical methods are introduced and then discussed in the context of the enzyme-immunoassay stability study from Chapter 4.

6.1 Transformations

When we conduct an experiment, we hope that the factors included in the design will explain the observed changes in process performance. In order to determine this, we fit a statistical model which can be used to predict values of the response at each of the design points. The statistical model makes certain assumptions about how each factor affects the response and about how the residuals (or random error) behave. Sometimes it may be necessary to analyze a transformation of the response variable, e.g., taking the natural logarithm of the response, in order to satisfy the assumptions of the statistical model. In this section, we discuss common transformations and how to select the best one.

Reasons for transformations

The assumptions implicit in fitting a statistical model to data from a fractional factorial or related experimental design are as follows (Weisberg, 1985):

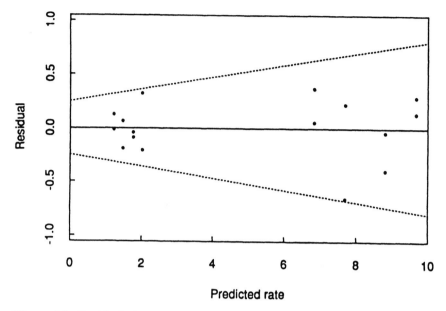

Figure 6.1. Residual plot showing nonconstant variances.

- the experimental factors and their two-factor interactions have a linear effect on the response
- the residuals (observed - predicted values) have a constant variance
- the residuals are symmetrically distributed about zero.

If one or more of these assumptions aren't satisfied it may be possible to find a transformation of the response which can correct the problem. In addition, it is possible that the use of the best transformation may result in a simpler statistical model.

Generally we can determine whether or not the assumptions of the statistical model are met by examining the errors or residual values. (Recall that a residual value is the difference between an observed response value and the value predicted from the statistical model.) The residuals are informative because they show how well the model represents the data.

When we choose a transformation, we are generally trying to improve the behavior of the residual values. We hope that a transformation which minimizes the residuals will also correct any problems with the distribution of the residuals; namely, nonconstant variance and lack of symmetry. Figure 6.1 shows the residuals versus predicted values from the statistical model fit to the example in Chapter 4. This plot clearly shows there is a problem with nonconstant variances of the residuals. (We added two lines to this plot to empha-size the pattern). This pattern, which shows that the variability of the residu-als increases as the response increases, is typical of data sets for which a transformation of the response will improve the fit of the model.

Standard transformations

The transformations which we use to improve model fit fall within the fami-ly of power transformations; that is, the response is raised to a power. The standard transformations from this family include the following:

- no transformation,
- square root,
- natural logarithm
- inverse square root, and
- reciprocal.

This family of power transformations can be expressed simply as y^λ, where, for example, $\lambda = 1/2$ represents the square root transformation, and we interpret $\lambda = 0$ as the natural logarithm.

These transformations are all especially useful for measurement data in the physical and biological sciences because they correct for right skewness (Carroll and Ruppert, 1988). Right skewness is a departure from symmetry which indicates that there are a number of values which are quite large in com-parison to the mean. They are also often useful in correcting for nonconstant variance.

6.2   Box-Cox family of transformations

Box and Cox (1964) proposed a family of power transformations which include the standard transformations described above. A generalization of their origi-nal transformation, which corrects for changes in scale induced by the transfor-mation, is as follows (see Weisberg, 1985):

$$y^{(\lambda)} = \frac{y^\lambda - 1}{\lambda * \dot{y}^{\lambda-1}} \qquad \text{if } \lambda \neq 0, \text{ and}$$

$$y^{(\lambda)} = \dot{y} * \ln(y) \qquad \text{if } \lambda = 0.$$

Here $\dot{y}$ is the geometric mean of the data; that is,

$$\dot{y} = (y_1 * y_2 * ... y_n)^{1/n} \quad \text{where each } y_i > 0.$$

The Box-Cox transformations have the following correspondence with the standard transformations identified above:

- $\lambda=1$; no transformation
- $\lambda=.5$; square root
- $\lambda=0$; natural logarithm
- $\lambda=-.5$; inverse square root
- $\lambda=-1$; reciprocal.

One way to determine which transformation, if any, should be applied is to fit the statistical model using a range of values for λ and then choose the value of λ which provides the best fit. By best fit, we mean the model which provides the smallest residual sum of squares, RSS_λ, for the transformed data $y^{(\lambda)}$. Because of the way the transformation is constructed, the values of RSS_λ can be directly compared (Cook and Weisberg, 1982).

Another, equivalent method is to calculate the log-likelihood function for each value of λ; namely,

$$L(\lambda) = - (n/2)*\ln(RSS_\lambda).$$

where n is the number of data values. The value of λ which minimizes RSS_λ also maximizes $L(\lambda)$. One advantage of the log-likelihood function is that an approximate confidence interval can be calculated for λ as follows:

$$\lambda: L(\lambda) > L(\hat{\lambda}) - (1/2)\chi^2(\alpha;1).$$

where $\hat{\lambda}$ is the value of λ which maximizes $L(\lambda)$ and $\chi^2(\alpha;1)$ is the α critical point of the chi-squared distribution with 1 degree of freedom.

A step-by-step procedure for selecting the best transformation is as follows:

- calculate $L(\lambda)$ for $\lambda=-1,+1$, by .1 (In some cases, it may be necessary to search beyond the range of -1 to +1 to find the best transformation.)
- identify the value of λ which maximizes $L(\lambda)$
- calculate the 95% confidence interval for λ
- plot $L(\lambda)$ versus λ and mark the confidence interval with a horizontal line
- use the standard transformation which falls inside the confidence interval.

If more than one standard transformation falls within the confidence interval, we would ordinarily choose the standard transformation closest to the value of

λ which maximizes $L(\lambda)$. Once a transformation has been chosen using this method, actual modeling can be done using the simpler form of the power transformation described in the previous section. A special program for carrying out this analysis is available as part of the SCA Statistical System (1986). It can also be readily implemented in most statistical packages which provide a programming environment.

Lambda plots

Our first objective in picking a transformation is to obtain a better fitting model. However, an added benefit of using an appropriate transformation may be a simplification of the model. One drawback of the standard Box-Cox analysis and subsequent plot of $L(\lambda)$ against λ is that we cannot see easily determine if the right transformation will result in a simpler model.

An alternative approach is to construct a lambda plot which provides more information about how the transformation changes the importance of various terms in the model. The lambda plot shows the t-values for all of model terms against the values of λ (Box and Fung, 1983 and 1986, Box, 1988). This method can be easily implemented using the SCA Statistical System (1986). In the next section, we apply this method to an example in order to gain a better understanding of how the choice of transformation can simplify the model.

A simple alternative method

One disadvantage of the Box-Cox method is that it is not readily available in commercial statistical software packages. A simple alternative, which is less powerful, but may still be useful, is to fit the model with each of the standard transformations and then pick the transformation which gives the best R-squared. The weakness of this method is that it doesn't provide as much help in determining which transformation provides the simplest model. The advantage of this method is that it can be easily carried out with most statistical software packages.

6.3   Example: Enzyme-immunoassay stability

In order to illustrate the use of transformations, we reconsider the immunoassay stability example discussed in Chapter 4. Recall that the experiment investigated the effects of five experimental factors (pH, thimerosal [thimer], gentamicin [gent], azide, and chelex) on the degradation rate of an enzyme-immunoassay component. The statistical analysis showed that the following effects were important: pH, thimer, pH*thimer, and pH*gent.

Transformation analysis

We already used the plot of the residuals versus predicted for this example (Figure 6.1) to show that the variability of the residuals increases as the response increases. Thus, we would like to select a transformation which will correct for this nonconstant variance. Hopefully, a simpler model will also result.

The reduced model which was used to generate Figure 6.1 included the terms pH, thimer, gent, pH*thimer, and pH*gent. We used the statistical package SCA (1986) to calculate the residual sum of squares (RSS_λ) and the t-values for each value of λ. These results are shown in Table 6.1.

Note that the minimum RSS_λ value (maximum log-likelihood) occurs when $\lambda=0.4$ (we have only evaluated the transformation at intervals of 0.10). Figure 6.2 shows the plot of $L(\lambda)$ $[=(16/2)*ln(RSS_\lambda)]$ versus λ. The value of $\chi2(.05,1)=3.84$, so the 95% confidence interval is

Table 6.1 Results of Box-Cox Power Transformation

			T-Values			
λ	RSS	pH	Gent	Thimer	pH*t	pH*g
-1.000	3.49	-22.8	-1.56	-5.17	-3.88	-2.06
-.900	2.99	-24.0	-1.51	-5.26	-3.75	-2.10
-.800	2.57	-25.4	-1.46	-5.35	-3.59	-2.16
-.700	2.22	-26.8	-1.40	-5.46	-3.40	-2.21
-.600	1.93	-28.3	-1.32	-5.58	-3.18	-2.28
-.500	1.68	-29.8	-1.23	-5.72	-2.93	-2.35
-.400	1.47	-31.5	-1.12	-5.87	-2.64	-2.42
-.300	1.30	-33.2	-1.00	-6.04	-2.31	-2.51
-.200	1.16	-34.9	-.852	-6.23	-1.93	-2.60
-.100	1.04	-36.7	-.686	-6.43	-1.51	-2.70
.000	0.95	-38.3	-.498	-6.64	-1.04	-2.80
.100	0.88	-39.9	-.289	-6.85	-.519	-2.90
.200	0.82	-41.3	-.060	-7.05	.037	-3.00
.300	0.79	-42.4	.184	-7.24	.625	-3.10
.400	0.77	-43.2	.440	-7.40	1.23	-3.18
.500	0.78	-43.6	.700	-7.53	1.84	-3.25
.600	0.80	-43.6	.956	-7.62	2.43	-3.30
.700	0.84	-43.2	1.20	-7.67	2.98	-3.33
.800	0.90	-42.5	1.43	-7.67	3.49	-3.35
.900	0.98	-41.4	1.64	-7.64	3.95	-3.35
1.000	1.10	-40.1	1.83	-7.59	4.35	-3.33

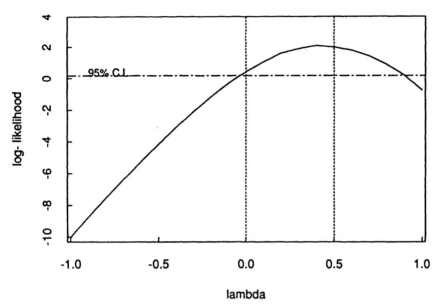

Figure 6.2. Plot of L(λ) versus λ.

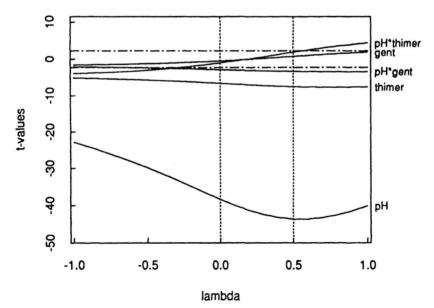

Figure 6.3. Lambda plot.

$$L(\lambda) > -8*\ln(.77) - (1/2)*3.84 = .171$$

Thus the horizontal line in Figure 6.2, which crosses the y-axis at .171, shows the 95% confidence interval for the true power transformation.

Both the square root and natural logarithm transformations fall within the 95% confidence interval (the corresponding values of $\lambda=.5$ and $\lambda=0$, respectively, are marked by vertical dotted lines). The square root transformation is closer to the maximum than in the logarithm, but we should examine the lambda plot before making a final choice of a transformation.

Lambda plot

The t-values for each of the five model terms is also shown in Table 6.1. These values were used to create Figure 6.3. The dashed horizontal lines represent the 95% confidence level for significant effects.

In this t-plot, we are primarily concerned with the of the choice of transformation on the pH*thimer interaction (pH*t in Figure 6.3). We are not concerned with pH, thimer, and pH*gent (pH*g) because they are each significant over the full range of transformations falling within the 95% confidence interval. The gent effect is not significant within this range.

There are two standard transformations which fall within the 95% confidence interval for λ (see Figure 6.2); namely, the natural logarithm and square root. The vertical dotted lines correspond to these two transformations. Note that the pH*thimer interaction does not have a significant t-value for either of these transformations. Thus, it appears that either transformation will lead not only to a better fitting model but also to a simpler model.

Picking the best transformation

In the absence of theoretical reasons for choosing between these two transformations, we can proceed to pick the best transformation based on our analysis so far. Thus, we choose the square root transformation because of it results in a somewhat better fitting model than the log transformation.

A better understanding of the pH*thimer interaction can be obtained from examining the square plots in Figures 6.5 and 6.6. Figure 6.5, based on the untransformed data, shows that the effect of thimer is much greater at the lower pH than at higher pH. Figure 6.5, based on the square root transformation, does not show this strong two-factor interaction. This suggests that thimerosal has closer to a proportional effect on degradation than an additive (linear) effect. (A strictly proportional effect would suggest the log transformation.) Note that the same effect is not observed for the pH*gent interaction.

The ANOVA table for the final model is shown in Table 6.2. The plot of residuals versus predicted values for this model is shown in Figure 6.6. Clearly

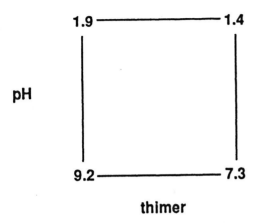

Figure 6.4 Square plot of pH and thimer for untransformed data.

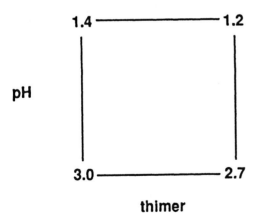

Figure 6.5 Square plot of pH and thimer for square root transformed data.

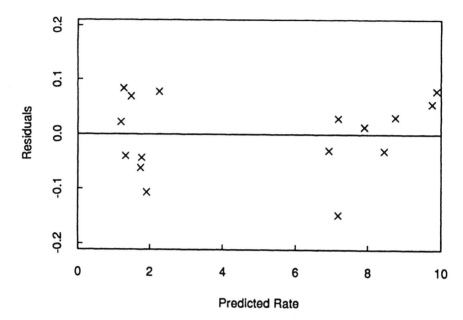

Figure 6.6. Residuals versus predicted for final, transformed model.

Table 6.2 **ANOVA for Model Without pH*thimer, Using Square Root**

R-Square 0.993
Adjusted R-Square 0.991
Residual Standard Error 0.081

Source	SS	df	MS	F-Ratio
Regression	10.60	4	2.65	405
Residual	0.07	11	0.007	
Adj. Total	10.67	15		

Effect	Coeff	Std Err	T-Value	P-Value
Intercept	2.07	0.0202	102	<.0001
pH	-0.80	0.0202	-39.6	<.0001
Thimer	-0.14	0.0202	-6.83	<.0001
Gent	-0.01	0.0202	-0.63	.54
pH*gent	-0.06	0.0202	-2.95	.01

Table 6.3 R-Squared Values for Various Transformations

| | R-Squared Values | |
Transformation	With pH*thimer	Without pH*thimer
Untransformed	99.4%	98.3%
Square root	99.5%	99.3%
Log	99.3%	99.3%
Inverse square root	98.9%	98.0%
Reciprocal	98.3%.	95.7%

this is a good model. There is no longer any pattern indicating nonconstant variance for the residuals.

Comparing R-squared values

In comparison to the more formal Box-Cox type of transformation analysis, we can also compare the R-squared values obtained by fitting the statistical model with the various transformations. Using the final model obtained in Chapter 4, which includes pH, thimer, gent, pH*thimer, and pH*gent, and the smaller model (without the pH*thimer interaction) suggested by the transformation analysis, the R-square values in Table 6.3 were obtained.

The differences in R-squared values among the transformations aren't very large when the extra term, pH*thimer, is included in the model. This is because the extra term partially compensates for being in the wrong metric (using the wrong transformation). However, the R-squared values fall off more quickly as we move away from the optimal transformation once the analysis is performed with the correct model. This R-squared analysis generally points toward the square root transformation, although it doesn't provide as much sensitivity as the Box-Cox type analysis for discriminating between the square root and log transformations.

6.4 Conclusion

In general the reasons to use a transformation of the response when fitting a statistical model are to

- allow the use of a simpler model.
- correct for nonconstant variance
- correct for skewed residuals

We described the use of a family of power transformations, which is generally referred to as a Box-Cox approach, to accomplish these objectives. Using this method we reanalyzed the data from the example of Chapter 4 to achieve a simpler, better fitting model with a more constant variance and better behaved residuals.

References and Bibliography

Box, G. E. P. (1988). Signal-to-noise ratios, performance criteria, and transformations. *Technometrics*, 30, 1-17.

Box, G. E. P. and D. R. Cox (1964). An analysis of transformations (with discussion). *J. Royal Statist. Soc. Ser. B*, 26, 211-246.

Box, G. E. P. and C. A. Fung (1983). Some considerations in estimating data transformations. MRC Report #2609, University of Wisconsin-Madison.

Box, G. E. P. and C. A. Fung (1986). Studies in quality improvement: Minimizing transmitted variation by parameter design. Report 8, University of Wisconsin-Madison, Center for Quality and Productivity Improvement (submitted to the *Journal of Quality Technology*).

Box, G. E. P., W. G. Hunter, and J. S. Hunter (1978). *Statistics for Experimenters*. New York: Wiley.

Carroll, R. J. and D. Ruppert (1988). *Transformation and Weighting in Regression*. New York: Chapman and Hall.

Cook, R. D. and S. Weisberg (1982). *Residuals and Influence in Regression*. New York: Chapman and Hall.

SCA Statistical Analysis System (1986). *Quality and Productivity Improvement Using the SCA Statistical System*. Scientific Computing Associates, P. O. Box 625, DeKalb, Illinois 60115.

Weisberg, S. (1985). *Applied Linear Regression*. 2nd edition. New York: Wiley.

Chapter 7

MIXED-LEVEL DESIGNS

In this chapter, we discuss the statistical analysis of fractional factorial and related designs in which one or more factors have three rather than two levels. These designs are useful because they allow the estimation of a curvature term. However, because not all factors have the same number of levels, as is true of a standard two-level fractional factorial design, some parts of the statistical analysis must be adjusted for differences in standard errors of the model effects. New aspects of the statistical analysis associated with these designs are introduced and then discussed in the context of a study to determine process variables which affect coating of a polyurethane surface with a bioactive material.

7.1 Detecting curvature

We are interested in curvature as an indication that the best response occurs in the interior of the experimental region as in Figure 7.1 This behavior cannot be represented by a linear change in the response between the low and high levels of the experimental factor. However, by adding a middle level for one or more factors and including corresponding curvature in the experimental model, we can detect such nonlinear behavior.

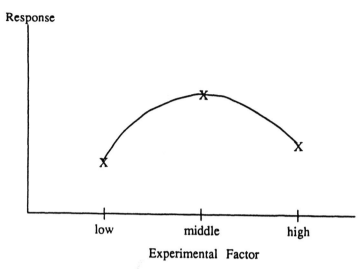

Figure 7.1 An illustration of curvature in the response.

Designs for detecting curvature

In order to detect curvature, we have already said that it is necessary to add a third level to one or more of the experimental factors. There are two basic ways to do this; namely, add center points to a standard two-level fractional factorial or select one of the mixed-level designs included in the Design Digest.

For example, the Design Digest recommends the use of three center points with design FF0516. As presented in Chapter 5, each center point is run at exactly the same conditions; namely, the middle level of each experimental factor. Contrast this design with MF0524 which allows only the first factor to have a middle level.

When using a fractional factorial design with center points, we get a generalized curvature term which cannot be tied to any single factor. However, the mixed-level designs allow the detection of nonlinear behavior in response to a specific three-level factor. A mixed-level design is preferable when there is prior information which suggests that a specific factor is likely to have its best setting in the middle of its limits. The use of center points is recommended when no prior judgement can be made as to which factor causes the nonlinear behavior. For example, center points are recommended when the high and low settings for each factor have been centered around the best settings obtained so far.

Adding a curvature term to the statistical model

For the statistical models of interest here, curvature can be represented as a squared term; that is, the expansion of the model from linear to quadratic. However, instead of just squaring the appropriate factor, we need to recode it in order to preserve the orthogonality of the design (independence properties) and to get the correct estimates of all the effects associated with the squared term.

The levels of a three-level factor are coded as -1, 0, and +1 for the statistical analysis. When this factor is squared, the levels become 0 and +1. However, we recode the curvature term so that the 0 becomes -1 and +1 becomes +.5. For example, if the three level factor is pH, then the curvature term (quadratic effect) is defined as follows:

$$pH2 = 1.5*pH^2 - 1.$$

It can be readily verified that when pH is -1 or +1, then pH2 is +.5 and when pH is 0, then pH2 is -1. With this coding any coefficient involving the squared term must be multiplied by 1.5, rather than 2, to convert it to an effect.

When the effect of pH2 is positive, the average response at the high level of pH2 (pH=-1 and +1) will be greater than the average response at the low level (pH=0). Thus, the response is lower in the middle. Conversely, if pH2 is negative, the response is higher in the middle. If we are trying to maximize the response in order to improve process performance, a negative value for the estimated effect of pH2 is desired.

7.2   Adjusting effects for differences in standard errors

The advantage of using a mixed-level design is that it is possible to detect curvature in the response function if it exists. The disadvantage is that mixed-level designs generally have slightly differing degrees of variability associated with some of the model effects. With most commercial statistical software packages, we can correct for these differences at appropriate places in the statistical analysis.

Designs with center points

When we fit a statistical model, as long as there are degrees of freedom left over for the error, the output of the analysis includes a value for the standard error for each effect. The standard error is similar to a standard deviation and measures the level of uncertainty associated with the estimated effect. The standard error of each effect is a function of the overall level of variability (the experimental error) and of the number of observations at each level of the corresponding factor.

In a standard fractional design, FF0516 for example, the estimate of each effect is the difference in averages between the 8 observations at the high setting and the 8 observations at the low settings. However, the effect due to curvature is the difference in averages between all 16 runs and the observations at the center point (3 center points are recommended by the Design Digest for FF0516). These differences between the number of observations averaged together to estimate the main effects, as opposed to the curvature, cause the related standard errors to be different.

Nonorthogonal designs

The standard fractional factorial designs discussed in earlier chapters are called orthogonal designs because all of the effects which they estimate are independent of each other. For example, a standard Resolution V design provides independent (orthogonal) estimates of all main effects and two-factor interactions whereas a standard Resolution III design provides independent (orthogonal) estimates only of the main effects.

One of the drawbacks of a standard orthogonal design is that it may require a larger sample size than really seems necessary to estimate the effects of interest. For example, to get a Resolution V design for six factors, a standard fractional factorial design requires a sample size of 32 (FF0632). In this case, there are only 22 effects which need to be estimated (6 main effects, 15 two-factor interactions, and the intercept). Thus, it is natural to wonder what we would have to give up in order to get a smaller sample size (say, closer to 22).

We have found that by allowing some correlation (lack of independence or partial confounding) among certain estimated effects, considerable savings in sample size can be achieved in comparison to standard orthogonal designs. For example, the design IF0624 is listed in the Design Digest as being *nearly* Resolution V. By *nearly*, we mean that there are limited correlations (partial confounding) among the estimable effects. In particular, although all of the main effects are independent of each other, not all of the main effects are completely independent of the two-factor interactions. For example, main effect B is correlated with the A*B interaction (+.33) and the E*F interaction (-.33). None of the correlations exceed +/-.33.

This correlation structure should not be a problem for any standard statistical software package which uses regression analysis to estimate effects. However, adjustments need to be made at a few points in the analysis to compensate for the differences in standard errors among the effects caused by the nearly orthogonal structure. These adjustments are described in the next subsection and they are illustrated with an example in the next section.

In addition to a series of irregular fractional factorial designs (IF series) with each factor at two levels, the Design Digest also includes a series of mixed-level fractional factorial designs (MF series) which have one factor at three levels. The design used in the example of Section 7.3, MF0524, is a member of this series. Complete listings of these designs are provided in the Design Digest. Designs for which there are correlations among some of the effects are listed as *nearly* Resolution IV or V, as appropriate.

Correcting for differences in standard errors

There are only two places in the statistical analysis that we need to correct for differences in standard errors among effects; namely, in the normal plot and in the active contrast plot. When we fit the statistical model, as long as there is at least one degree of freedom that is left over for the error, the output includes an estimated standard error for each effect or coefficient. The assumptions of the normal plot and the active contrast plot will be met if we simply divide the estimated effects by their standard errors before plotting them or using them in the calculation of the active contrast plot. We can also use these standardized values in the Pareto chart if we wish. (Note that the output of most regression programs will be coefficients and standard errors of the coefficients. These may be converted to effects and standard errors of the effects by multiplying by two in each case.)

7.3   Example: Coating a polyurethane surface with a bioactive material

In the following example, we use a mixed-level design, MF0524, to identify the important factors affecting the coating a polyurethane surface with a bioactive material.

Problem description

In this process, we wish to coat a polyurethane surface with a bioactive material. The coating process may consist of up to three steps. The main stage of the process consists of immersion in a bath containing the bioactive material. The usefulness of pretreatment and posttreatment baths are also being considered. The response of interest (bound) is a reading of the average density of the bioactive material remaining on the surface. We wish to maximize the response.

In the first step, the samples are submerged in a pretreatment bath containing a chemical which modifies the surface of the polyurethane. It is believed that the modification improves the ability of the bioactive material to bind to the surface. However, this chemical has been previously associated with toxic effects, and it is desired to minimize its use if possible.

In the second stage of the process, the bioactive material is also applied by submerging the samples in a bath. The following variables are of interest: the concentration of the bioactive material in the bath (conc), the time (in minutes) of the bath, and the temperature (temp) of the bath.

The last step of the process consists of a posttreatment bath (post).This bath contains a biological fixative which is believed to fix the bioactive material to the polyurethane surface.

It is thought that temperature may have a nonlinear effect on the amount of bioactive material bound. Thus, temperature will be considered at high, middle, and low levels. This allows the investigation of a wider range of temperatures than would ordinarily be considered in a two-level design. The best response may occur at the middle temperature.

The other factors are considered at two-levels each. For the pretreatment bath, the low and high levels represent concentrations of the active chemical in the bath. For the posttreatment baths, the low level represents no bath and the high level represents the use of a bath at a previously determined concentration of the active ingredient.

Experimental design

The experimental design selected was MF0524. The experimental design and resulting data are shown in Table 7.1. Note that the factor temp has three levels. Thus, we will need to adjust for differences in standard errors of the effects before constructing the normal and active contrast plots.

Referring to MF0524 in the Design Digest, we see that it is a 1/2 fraction of a $3*2^4$ factorial design. It is a nearly Resolution V design. We can estimate all main effects, all two-factor interactions, a squared term for temp, and the interactions between the other main effects and the squared term. No center points were run, so we can not estimate a generalized curvature term.

Experimental results

The best run overall is run #8 followed by run #24. Runs #12 and #14 also look quite good. A "pick the winner" strategy based on run #8 would select the higher concentration in the pretreatment bath, high time, high concentration, low temperature, and the posttreatment bath as the best set of conditions. A review of all four good runs shows that high concentration and the posttreatment bath are common settings. We suspect that these will be our most important factors.

The statistical analysis should provide more reliable and quantitative information about the relative importance of the various experimental factors and about their best settings. We are also interested in identifying trade-offs

Table 7.1   Experimental Worksheet for Coating Study

	Experimental Factors					Response
Run*	Temp	Prep	Time	Conc	Post	Bound
1	-1	-1	-1	-1	-1	3.8
2	-1	-1	-1	1	1	10.4
3	-1	-1	1	-1	1	12.8
4	-1	-1	1	1	-1	8.7
5	-1	1	-1	-1	1	5.1
6	-1	1	-1	1	-1	6.4
7	-1	1	1	-1	-1	3.8
8	-1	1	1	1	1	24.2
9	0	-1	-1	-1	1	7.6
10	0	-1	-1	1	-1	3.9
11	0	-1	1	-1	-1	4.3
12	0	-1	1	1	1	17.2
13	0	1	-1	-1	-1	3.5
14	0	1	-1	1	1	17.4
15	0	1	1	-1	1	8.2
16	0	1	1	1	-1	9.1
17	1	-1	-1	-1	-1	5.3
18	1	-1	-1	1	1	12.1
19	1	-1	1	-1	1	9.1
20	1	-1	1	1	-1	12.0
21	1	1	-1	-1	1	4.2
22	1	1	-1	1	-1	10.8
23	1	1	1	-1	-1	7.3
24	1	1	1	1	1	20.0

* Runs were conducted in randomized order to guard against systematic bias.

among the factors as we would like to reduce the concentration of the active chemical in the pretreatment bath if possible.

Fitting a preliminary statistical model

Before fitting the first model, we create a squared term to represent possible curvature in the three-level factor temp. We let temp2=1.5*temp^2-1. Since we are trying to maximize the response, temp2<0 will indicate that the best settings are at the middle level of temp.

Now we can fit a statistical model with the following terms:

response variable:

- bound

main effects:

- temp, prep, time, conc, post

two-factor interactions:

- temp*prep, temp*time, temp*conc, temp*post
- prep*time, prep*conc, prep*post
- time*conc, time*post
- conc*post

squared term:

- temp2

Table 7.2   Estimated Effects from Untransformed Model

Effect Name	Estimate Effect	Standard Error of Effect
Temp	0.70	0.834
Prep	1.05	0.681
Time	3.85	0.681
Conc	6.41	0.681
Post	5.79	0.681
Temp*prep	-0.03	0.834
Temp*time	-1.00	0.834
Temp*conc	0.57	0.834
Temp*post	-2.48	0.834
Prep*time	-0.60	0.722
Prep*conc	2.51	0.722
Prep*post	0.22	0.722
Time*conc	1.12	0.722
Time*post	1.10	0.722
Conc*post	2.82	0.722
Temp2	0.84	0.722
Prep*temp2	-0.18	0.722
Time*temp2	1.70	0.722
Conc*temp2	0.31	0.722
Post*temp2	-1.17	0.722

interactions with squared term;

- prep*temp2, time*temp2, conc*temp2, post*temp2

Including the intercept, there are 21 model terms (the intercept, 5 main effects, 10 two-factor interactions, the squared term, and 4 interactions with squared term). Since there are 24 observations, there are 3 degrees of freedom left for the error. The estimated effects and their standard errors are given in Table 7.2. A Pareto chart of these effects is shown in Figure 7.2.

Figure 7.2 shows that conc and post are the most important effects. It looks like time and the two-factor interactions for conc*post, prep*conc and temp*post are probably also important effects. The time*temp2 interaction is in a questionable range. Other important effects seem unlikely.

Later in this chapter, we show how to adjust the normal plots and active contrast plots for differences in standard errors among the estimated effects (see Table 7.2). However, we did not make this adjustment for the Pareto chart, because it seems important to preserve its original interpretation (each bar shows the difference in average response between the high and low settings for an effect). Omitting this adjustment is not likely to result in errors of interpretation as long as the differences in standard errors are not too great. This assumption is reasonable for the designs included in the Design Digest.

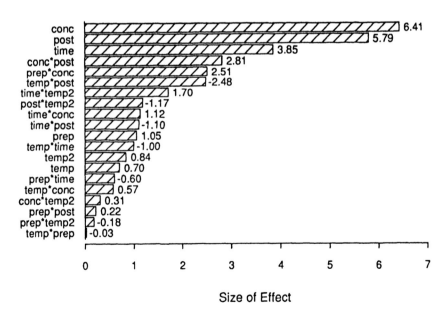

Figure 7.2 Pareto chart for coating study, untransformed data.

Comments on the model

The statistical model confirms our guess (based on examination of individual runs) that concentration and the posttreatment bath are the most important factors. They each have a positive effect so are best at their high levels. The next most important effects are time, conc*post, prep*conc, and temp*post. As a rough cut, conc and post are the "vital few", time through temp*post or possibly time*temp2 are the "statistical in-between" and the rest are probably among the "trivial many." The quadratic effect of temperature (temp2) seems to be rather small and in the opposite direction from that expected.

7.4   Selecting the best transformation

In the previous chapter, we saw that the proper choice of a transformation can result in a better model. Therefore, we perform a transformation analysis (as in Chapter 6) to identify the best transformation and determine its effect.

Transformation analysis

The plot of the log-likelihood function versus λ is shown in Figure 7.3. The maximum value of $L(\lambda)$ occurs at approximately $\lambda=.3$. The 95% confidence limit for the best transformation is shown by the horizontal line. The confidence interval contains both $\lambda=0$ and $\lambda=0.5$. We are inclined to use the square root transformation since $\lambda=0.5$ is closer to the maximum.

Before making a final choice of λ, we should examine the lambda plot of t-values. This plot is shown in Figure 7.4. The dashed horizontal lines show the 95% confidence levels for significant t-values. The vertical lines, drawn at $\lambda=0.0$ and $\lambda=0.5$, help us identify the significant t-values with the natural logarithm and square root transformations, respectively.

There is considerable room for debate as to whether the square root transformation or the logarithmic transformation is better. On the one hand, the square root transformation will give a better fit (although both are statistically reasonable). On the other hand, the logarithm gives a simpler model; that is, the two-factor interaction post*conc becomes nonsignificant with the log transform which suggests that post and conc have proportional effects on the amount of bioactive material bound. For this example, we feel that it is reasonable to go with the better fitting model. Hence, we use the square root of the amount of bioactive material bound as the response in the rest of the analysis.

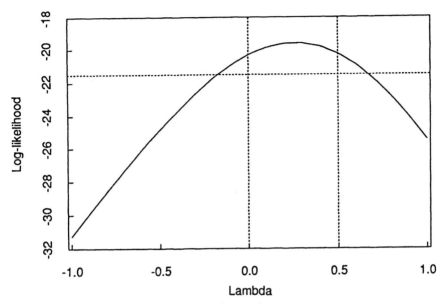

Figure 7.3 Plot of L(λ) versus λ for coating study.

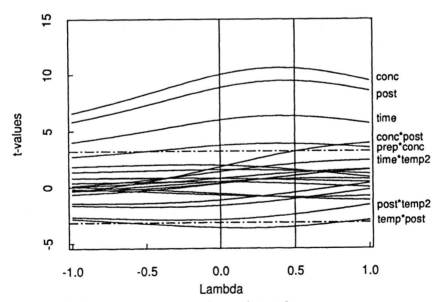

Figure 7.4. Lambda plot of t-values for coating study.

Table 7.3   Estimated Effects from Square Root Transformed Model

Effect Name	Estimated Effect	Standard Error of Effect
Temp	0.180	0.118
Prep	0.096	0.096
Time	0.607	0.096
Conc	1.019	0.096
Post	0.907	0.096
Temp*prep	0.032	0.118
Temp*time	-0.116	0.118
Temp*conc	0.088	0.118
Temp*post	-0.420	0.118
Prep*time	-0.093	0.102
Prep*conc	0.394	0.102
Prep*post	-0.050	0.102
Time*conc	0.129	0.102
Time*post	0.104	0.102
Conc*post	0.324	0.102
Temp2	0.146	0.102
Prep*temp2	-0.110	0.102
Time*temp2	0.210	0.102
Conc*temp2	0.062	0.102
Post*temp2	-0.250	0.102

7.5   Graphical Identification of important factors

We are now ready to apply the methods of Chapter 4 to identify important factors. The estimates of effects for the square root transformation are shown in Table 7.3, and the Pareto chart is shown in Figure 7.5.

Normal plot

One of the assumptions of the normal plot is that all of the effects should have the same standard error. This is not the case with design MF0524 (see Tables 7.2 and 7.3). This problem can be corrected by dividing each effect by its standard error before creating the normal plot. The resulting plot is shown in Figure 7.6. The effects of conc, post and time fall well off a hypothetical line through the middle of the effects. The effects of conc*post and prep*conc also fall somewhat below the line on the right hand side of the plot. The effects of temp*post and post*temp2 seem to be above a line of the right hand side of the plot. No other possibly significant effects are evident.

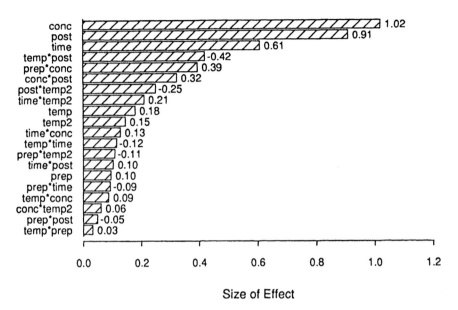

Figure 7.5 Pareto chart for square root transformed model.

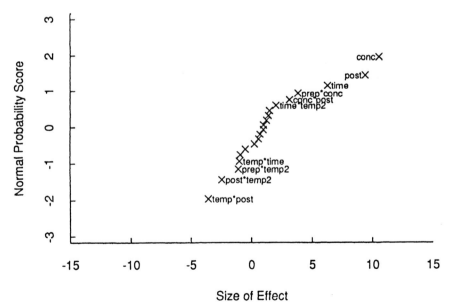

Figure 7.6 Normal plot for square root transformed model.

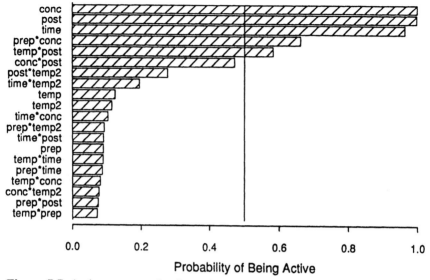

Figure 7.7 Active contrast plot for square root transformed model.

Active contrast plot

The active contrast plot (Figure 7.7) should also be constructed using standardized effects. This plot was constructed using a prior probability for active effects of 0.3 This presupposes that up to 30% or 6 of the 20 effects may be significant. The scale factor for this plot was 5.

Figure 7.7 shows that time, conc, post, prep*conc and temp*post are clearly important effects. While conc*post does not quite fall above the .5 cutoff level, it seems large enough so we should consider it as a candidate term in the reduced model. It can be deleted from the reduced model later if its t-value is not significant. The post*temp2 interaction is not important. Note that the active contrast plot seems to provide a more sensitive analysis than the normal plot. Our practical experience generally confirms this observation.

Cube and square plots

Since there are three important main effects, we first consider the cube plot of post, conc, and time. The appropriate average of the square root of bound values is plotted at each vertex. This is shown in Figure 7.8. In this cube plot, we see that more bioactive material is bound to the polyurethane surface at higher concentrations, longer time, and with the posttreatment bath. This is the upper right hand corner of the cube.

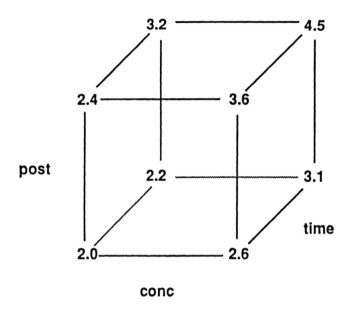

Figure 7.8 Cube plot of post, conc, and time for square root transformed data.

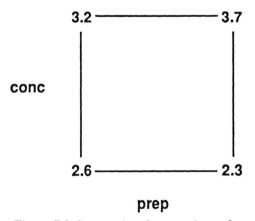

Figure 7.9 Square plot of conc and prep for square root transformed data.

The effect of concentration is evident as we go from the left face to the right face of the cube (low to high concentration); the amount bound increases along each edge. Similarly, the amount bound increases along each edge from the bottom face to the top face of the cube (bottom face = without posttreatment bath, top face = with posttreatment bath) and from the front face to the back face (going from low to high settings of time).

The effect of the conc*post interaction can be seen on either the front or back face of the cube. For example, on the front face, the addition of a posttreatment bath has a much greater effect on the amount of bioactive material bound at the high concentration than compared to the low concentration.

In order to clarify the effect of the conc*prep interaction, we examine the square plot for conc and prep shown in Figure 7.9. It appears that prep has a negative effect at the low concentration and a positive effect at the high concentration.

Finally, we wish to examine the temp and the temp*post interaction. First, consider the interaction plot in Figure 7.10. This plot shows the average response at each combination of temp and post. The upper line in Figure 7.10 shows the average responses at the high level of post and the lower lines corresponds to the low level of post. The averages for temp=0 are neither higher nor lower than the averages for temp=-1 and temp=+1. This confirms the lack of significance of the squared term for temperature (temp2). It also suggests that the only conditions under which temperature seems to matter is at the combination of no posttreatment bath (post=-1) and highest temperature (temp=+1).

More information about the temp*post interaction is provided by the square plot for temp and post, which is shown in Figure 7.11. Note that the temp*post interaction actually does not include any information from the values when temp=0. (These latter values determine the temp2*post interaction, which did not seem to be important.) The square plot shows clearly that a posttreatment bath has the greatest effect when the treatment bath is run at the low temperature.

Summary of results of graphical analysis

The graphical analysis suggests that the following effects are important:

main effects

• conc, post, time

two-factor interactions

• conc*post, prep*conc, temp*post

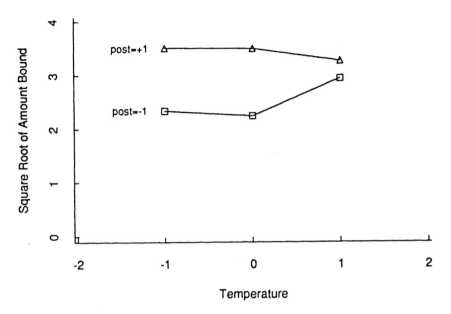

Figure 7.10 Interaction plot for post*temp, square root transformed data.

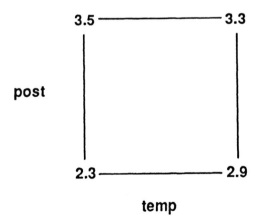

Figure 7.11 Square plot of post*temp, square root transformed data.

It also appears that there is no curvature associated with the three levels of temperature. The main effects prep and temp aren't important by themselves, but because of their interactions with important main effects, we will include them in the reduced model.

7.6   Evaluation of the reduced model

As described above, the reduced model has 5 main effects and 3 two-factor interactions. We use the square root of the amount of bioactive material bound as the response. In this section, we fit this model and evaluate the fit.

ANOVA results

The ANOVA results are presented in Table 7.4. First we note that the R-squared value of 92.9% (.929) and the adjusted R-square value of .891 are very high. The residual standard error or estimated experimental error at 0.28 seems reasonable based on prior experience with this process. Recall that we

Table 7.4   ANOVA Table for Reduced Model Using Square Root

R-Square	0.929
Adjusted R-Square	0.891
Residual Standard Error	0.280

Source	SS	df	MS	F-Ratio
Regression	15.881	8	1.985	24.537
Residual	.214	15	0.081	
Adj. Total	17.094	23		

Effect	Coeff.	Std Err	T-Value	P-Value
Intercept	2.96	0.058	50.9	<.0001
Temp	0.09	0.071	1.26	.23
Prep	0.05	0.058	0.82	.43
Time	0.30	0.058	5.22	<.001
Conc	0.51	0.058	8.77	<.0001
Post	0.45	0.058	7.81	<.0001
Temp*post	-0.21	0.071	-2.94	.01
Prep*conc	0.21	0.058	3.68	<.01
Conc*post	0.15	0.058	2.51	.02

can calculate the R-squared value by dividing the regression sum of squares by the adjusted total sum of squares; that is, 15.881/17.094=.929.

Next we consider the coefficients, t-values and p-values presented in the lower half of Table 7.4. Each of the three main effects is very highly significant. The three two-factor interactions are each very significant. We are not particularly concerned with the main effects of temp and prep which we added for completeness.

Graphical evaluation of reduced model

Four plots will be constructed to assist in evaluating the fit of the reduced model; namely, the predicted versus observed plot (Figure 7.12), the residuals versus predicted plot (Figure 7.13), the residuals versus run order (Figure 7.14) and the normal plot of residuals (Figure 7.15).

The predicted versus observed plot shows that the points cluster satisfactorily about the 45 degree line (Figure 7.12). There don't appear to be any patterns indicating shortcomings in the model. The plot of residual values against predicted values should show random scatter (Figure 7.13). We don't expect to see a pattern indicating nonconstant variances since we have already done a transformation analysis. The plot in Figure 7.13 looks quite satisfactory.

The plot of residuals versus run order is intended to help identify the effects of any systematic bias; for example drift in instrument readings. No special patterns are obvious in Figure 7.14. Finally, the normal plot of residuals evaluates the behavior of the residuals in comparison to a normal distribution. We watch especially for an "S" shaped curve or for one or more points which fall far from a line passing through the center of the residuals. The four most extreme negative residuals seem somewhat detached from the body of the residuals. However, this doesn't represent a serious departure from normality.

Thus, all of the graphical analysis confirms the satisfactory fit of the reduced model. That is, the model quite adequately explains variability in the amount of bioactive material bound to the polyurethane surface.

7.7   Interpretation and comments

Our analysis indicated that there were six important effects to consider in trying to improve coating. Three of these were main effects and three were two-factor interactions. In order to improve the fit of the model we used a square root transformation of the response. Although this design, MF0524, allows the estimation of a curvature term for the factor temp, this effect was not important.

Through significant main effects and/or two-factor interactions, each of the experimental factors in this experiment has an important effect on the perfor-

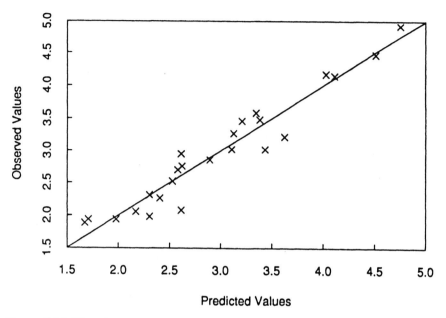

Figure 7.12 Plot of predicted values versus observed values, reduced model.

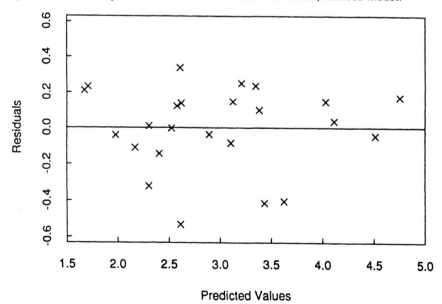

Figure 7.13 Plot of residual values versus predicted values, reduced model.

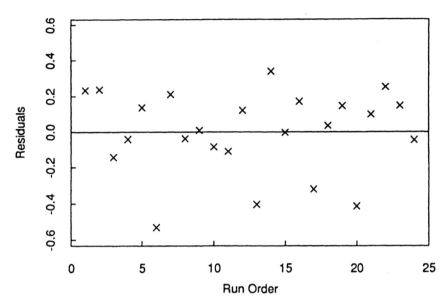

Figure 7.14 Plot of residual values versus run order, reduced model.

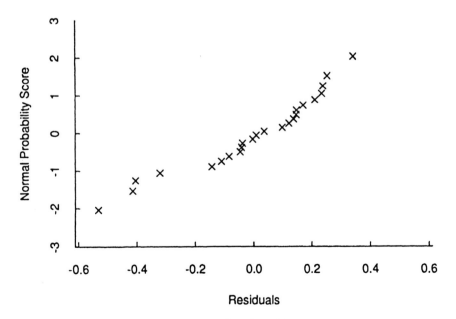

Figure 7.15 Normal plot of residuals, reduced model.

mance of the coating process. In order to improve the coating, the statistical analysis suggests that we should do the following:

- use the low concentration of the active chemical in the pretreatment bath (this is because of the significance of the prep*concentration interaction).
- use a treatment bath with the high concentration of bioactive material (main effect of conc), for the longer time (main effect of time), at the lower temperature (temp*post interaction).
- use the posttreatment bath.

Note that these recommended conditions concur with the results of the best run, #8. The next best run was run #24. We believe that this run had a lower yield because it was conducted at high temperature. Runs #12 and #14 apparently did well because they had high settings of concentration, included the posttreatment bath, and either had time or prep at its best setting. Each of these runs was at the intermediate temperature which did not significantly affect coating (recall Figure 7.10).

One of the concerns that we mentioned at the beginning of this chapter was that the pretreatment bath was thought to have possibly toxic side effects. If we wished to design a process with acceptable yield but which minimized the concentration of the active chemical in the pretreatment bath, we could model it after run #12 with the exception of using the low temperature.

If we review Table 7.4, which presented the coefficient estimates for the reduced model, we can quantify the expected decrease in yield associated with reducing the concentration in pretreatment bath. In particular, if prep is set to its low level, and all other factors are set at their best settings, the predicted yield is (in the square root scale) $2.96 + 0.05*(-1) + 0.30*(+1) +0.51*(+1) + 0.09*(-1) + 0.21*(-1*+1) +0.15*(+1*+1) - 0.21*(-1*+1) = 4.23$ units. In comparison, with prep at its high level, the predicted yield is $2.96 + 0.05*(+1) + 0.30*(+1) +0.51*(+1) + 0.09*(-1) + 0.21*(+1*+1) +0.15*(+1*+1) - 0.21*(+1*+1) = 4.75$ units. If we square these predicted values to convert back to original units, we predict 17.9 (with the lower pretreatment bath concentration) versus 22.6 (with higher pretreatment bath concentration).

Although the concentration of the active chemical in the pretreatment bath significantly effects coating, we are able to use the statistical model to predict how large an effect it has on the process yield. Given this information, it is possible to make a decision, which is based on good information, whether or not we can afford the trade-off between lower toxicity and higher yield which is associated with reducing the concentration of the active chemical in the pretreatment bath.

7.8   Conclusions

Good progress was made toward optimizing the coating process. This experiment identified the important factors affecting process yield and the suggested changes in the factor settings to improve the yield. A second experiment might attempt further improvements by considering the conditions of the posttreatment bath (for example, concentration, temperature, time, etc.). The second experiment might also attempt to further quantify trade-offs between lowered toxicity and reduced yield due to reducing the concentration of the active chemical in the pretreatment bath. Although there was strong enough belief prior to conducting the experiment that there might be a curvature effect due to temperature, this was not confirmed by the experimental results.

This experiment was designed, conducted and analyzed with the following considerations in mind:

- small, efficient experimental designs can be used to generate information-rich data
- clear signal designs guard against interactions among variables
- statistical graphics and analysis separate the signals from the noise - signals which are large in comparison to the noise are associated with important process variables
- a series of small experiments can be used to solve a complex biological problem
- appropriate experiments are conducted in the following sequence: screening, optimization, and verification - in this case a screening experiment
- a "Stop, Look, and Listen" approach to data collection and analysis leads to productive results.

References and Bibliography

Addelman, S. (1961). Irregular fractions of the 2^n factorial experiments. *Technometrics*, 3, 479-496.

Addelman, S. (1962). Orthogonal main-effect plans for asymmetrical factorial experiments. *Technometrics*, 4, 21-46.

Box, G. E. P. (1988). Signal-to-noise ratios, performance criteria, and transformations. *Technometrics*, 30, 1-17.

Box, G. E. P., W. G. Hunter, and J. S. Hunter (1978). *Statistics for Experimenters*. New York: Wiley.

Daniel, C. (1976). *Applications of Statistics to Industrial Experimentation*. New York: Wiley.

Dey, A. (1985). *Orthogonal Fractional Factorial Designs*. New York: Wiley.

Ghosh S. (1987). Non-orthogonal designs for measuring dispersion effects in sequential factor screening experiments using search linear models. *Communications in Statistics - Theory and Methods*, **10**, 2839-2850.

Hahn, G. J. and S. S. Shapiro (1966). A catalog and computer program for the design and analysis of orthogonal symmetric and asymmetric fractional factorial experiments. Report No. 66-C-165, General Electric Research and Development Center, Schenectady, New York.

John, P. W. M. (1969). Some non-orthogonal fractions of 2^n designs. *J. Royal Statistical Society, Series B*, **31**, 270-275.

Liddle, R. F. and P. D. Haaland (1988). Efficient nonstandard screening designs. Presented at the Annual Meetings of the American Statistical Association, New Orleans, LA.

Margolin, B. H. (1968). Orthogonal main effect 2^n*3^m designs and two factor interaction aliasing. *Technometrics*, **11**, 431-444.

Margolin, B. H. (1969). Orthogonal main effect plans permitting estimation of all two factor interactions for the 2^n*3^m series of designs. *Technometrics*, **11**, 431-444.

Ross, P. J. (1988). *Taguchi Techniques for Quality Engineering*. New York: McGraw-Hill.

Taguchi, G. (1986). *Introduction to Quality Engineering : Designing Quality into Products and Processes*. Tokyo: Asian Productivity Organization.

Taguchi, G. (1987). *System of Experimental Design*. Vol. 1 and 2. New York: Unipub.

Taguchi, G. and Y. Wu (1980). *Introduction to Off-line Quality Control Systems*. Tokyo: Central Japan Quality Control Association.

Chapter 8

STRATEGIES FOR EXPERIMENTERS

In this chapter, we describe iterative strategies which are useful for solving empirical problems. These strategies help us to focus on the important parts of the problem, identify important factors and improve process performance. The use of such strategies is particularly effective when a series of small experiments is used to solve a complex problem.

8.1 Overview

In preceding chapters, we discussed a number of statistical tools; for example,

- experimental design,
- statistical analysis,
- model building,
- statistical graphics,
- numerical optimization, and
- verification.

These tools are most useful when they are applied within an effective strategy for problem solving. For example, even the best statistical analysis will not identify an important factor which was left out of the experiment.

149

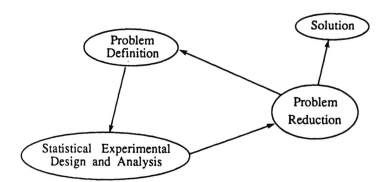

Figure 8.1 The problem solving strategy is iteratively applied until the problem is solved.

The first component of a successful strategy is problem definition; that is, we must be sure we are working on the right problem. The second part of a successful strategy involves formulating the problem so that we can systematically apply the methods of statistical experimental design and analysis. The final component of a successful strategy involves problem reduction; that is, identifying the important factors, eliminating unimportant factors, and moving closer to the solution. This strategy is typically applied in an iterative fashion (Figure 8.1) until a solution is reached.

Strategies for experimentation

The following list provides a summary of the key elements of the strategies we use for solving empirical problems:

Problem Definition:

- Clarify the experimental objectives.
- Identify response variables.
- Identify potential experimental factors.

Collecting and Analyzing the Data:

- Identify and control sources of experimental error.
- Select factors to include in the first experiment.
- Choose high and low settings for each factor.
- Select the statistical experimental design.

* Run the experiment and analyze the results.

Problem Reduction

* Identify important factors.
* Set qualitative factors at their best settings.
* Fix unimportant factors at economical levels.
* Pick levels for the important factors which should result in improved process performance in the next experiment.

In the rest of this chapter, we discuss these strategies for empirical problem solving.

8.2   Problem definition

The purpose of the problem definition strategies is to make sure that we are solving the right problem.

Experimental objectives

The first step in solving an empirical problem is to clearly define the experimental objectives. These objectives provide the criteria which determine when the problem has been solved. Having well defined objectives

* helps focus on the important aspects of the problem and
* makes sure we solve the right problem.

It is useful to think of experimental objectives in the context of the types of errors we can make in trying to achieve them. Most statistical texts discuss two types of error. For example, if we say we a particular process variable has an important effect on process performance when it really doesn't, we are making a Type I (alpha) error. If we say that a process variable isn't important when it really is, we are making a Type II (beta) error.

However, in problem definition, we are primarily concerned with avoiding *Type III* errors. Leaving out an important factor from the experimental design is an obvious way to commit a Type III error. Choosing the wrong measure of process performance can also lead a Type III error. For example, suppose we maximized antibody yield, but the purity declined so much that the antibody could not be economically purified.

It is important to remember that the experimental objectives and the criteria for evaluating them may be politically determined. Clear objectives and good politics can often be achieved by forming a project team consisting of research scientists, technicians, and managers responsible for the project. An integrated team like this encourages successful communication and allows all interested parties to "buy in" to the experimentation strategy.

Identify response variables

A response is a variable which is a measure of the outcome of an experiment. The value of the response(s) is used to determine whether or not the experimental objectives have been met. Picking the right responses is a crucial condition for successful empirical problem solving.

For each response, a criterion is applied in order to decide whether or not experimental objectives have been met. For example,

* maximize the yield from an antibody culture system,
* false positives from an screening test should be less than 10%,
* the average percent cell viability should be at least 80%,
* the zero binding reading should be at least .050 after one week of storage at 30 degrees C,
* the cost of materials used should be minimized,
* the slope of the standard curve should be as large as possible.

Ideally, responses should be easily and accurately measured. However, responses are not restricted to characteristics which may be easily measured. For example, a response may be qualitative (how blue is the colored dot), dichotomous (yes/no) or nominal (ordered categories such as "worse", "same", "a little better", "a lot better").

A good approach for qualitative variables is to look at all the samples at the same time and order them according to the qualitative response. Each sample can then be assigned a rank (from 1 to n) and the ranks can be analyzed. A dichotomous response might be whether an assay shows a false positive reading (1) or not (0). In this case 0's and 1's can be used as values for the response variable. In the sensitivity analysis of an assay, samples are often prepared at specified dilutions. The value of the highest dilution at which a positive result was observed can be used as the response. We have had good luck analyzing these types of responses. However, more information is usually obtained if a measurement can be made.

One way in which a series of experiments can get off track is if the primary response or the criterion applied to the response is changed after the experiments have begun. If the primary response is changed after the experiments are over, then it is possible that the wrong problem was solved. A more difficult problem is concerns the trade-offs required when the conditions which satisfied the criteria for one response don't satisfy the criteria for a second important response. We will see an example of how to deal with this problem in Chapter 9.

Identify experimental factors

Experimental factors (independent variables) are the process variables or conditions which the experimenter controls during an experiment; for example:

- the temperature at which a reaction takes place,
- the time allowed for a reaction,
- the pH of the buffer, etc.

Factors such as time, temperature, concentration, etc. are continuous variables. Factors such as number of washes, number of injections, etc. which take on discrete values are not a problem as long as we are careful about how we interpret results such as "the optimal setting is at 1.2 washes".

Qualitative variables such as type of buffer salts, type of antigen, etc., are easy to use as long as we are willing to consider only two or three values or levels at a time in a single experiment. (It is possible to modify fractional factorial designs to include four-level factors, but the results are difficult to interpret.) Designs of the type MF0524 are useful for these experiments. Additional small designs for three-level factors can be found in Taguchi (1986).

In addition to process variables, experimental factors might include the following: raw materials (for example, lot or brand), lab technician, supplier of antigen or reagent, etc. These would be used as experimental factors rather than blocking factors when the experimental objectives included specifically determining the effects of these factors on process performance.

An important experimental factor which has been left out of an experimental design is sometimes called a *lurking variable*. (This term was coined by G. E. P. Box. For more information, see Joiner, 1981.) If this important variable changes during the course of the experiment, there will be unexplained large jumps in the process performance. If the lurking variable changes in-between experiments, there will be an unexplained inconsistency in results between the two experiments. In either case, careful consideration should be given to potential experimental factors early in the problem definition process.

8.3   Collecting and analyzing data

Statistical experimental design and analysis provides a systematic framework in which we collect and interpret data which is subject to uncertainty (i.e., noise). In this section, we are not concerned with how to design an experiment or analyze data, both of which we have discussed at length earlier. Rather we are concerned with some issues in formulating the design and analysis.

Sources of experimental error

An important question scientists may ask is when should you begin to use statistical methods. Fractional factorial and related designs, which we use for screening experiments, require a certain level of control or reproducibility in the experimental error.

The need for control of experimental error does not imply that these methods are not good when there are high levels of noise in the data. In Chapter 5, we discussed power and sample size. We pointed out that for signals and noise of given magnitude, the chance of detecting the signal is a function of the sample size. In fact, statistics is particularly well suited to reasoning in the presence of this type of noise.

In controlling sources of experimental error, we mean assignable causes of experimental error. For example, suppose that a sample reagent varied in reactivity from lot-to-lot. This would contribute to uncontrolled variability if different lots of reagent were used to prepare samples and reagent were not used as a blocking variable. A similar argument would apply to having technicians of differing ability prepare or measure samples. The scientist should have a fairly high degree of confidence that secondary factors which affect process performance will not change during the course of an experiment.

Selecting experimental factors

The list of all process variables which might affect the response may be quite long. This is a problem because experimental effort increases quickly as more factors are included in the design.

A review of the list of potential factors may identify some factors which are impractical to include in the experiment. For example, in the ascites optimization problem discussed in Chapter 2, the age of the mice was not investigated because of the high cost of keeping mice while they age. It may also be necessary to omit other factors which have a low probability of being active.

On the other hand, factors which are left out of the experiment may later show up as lurking variables and contribute to confusion regarding the true causes of changes in process performance. Also, the predictions of process performance may not be valid if the setting of the important factor needs to be changed later. Leaving out an important factor may also make it impossible to achieve desired improvements in process performance.

The most effective approach requires a number of subjective judgements based on the researcher's knowledge of the process being studied. We suggest starting with an exhaustive list of factors, identifying those which are likely to affect process performance, including as many factors as possible in the first experiment, and eliminating factors from subsequent experiments based on a problem reduction strategy.

Choosing factor settings

In order to get started, we need to understand the feasible range of values for each factor. This range is determined by the highest and lowest possible settings for each factor at which the process will still work. The researcher may also have preferences with regard to factor settings; for example, the pH shouldn't be above 7.7 or while some antimicrobial agents must be added to the storage buffer the lowest possible levels are preferred. The feasible range and preferred values assist the researcher in picking the high and low levels to be used in the experiment.

The first experimental region is chosen so that there is likely to be a difference observed in the responses between the two levels of each factor. Depending on the specific factor, you may want to choose the high and low settings so that they straddle the best initial guess or you may want to choose two competing values. Usually, it is a good idea not to set either the high or low limits at the extremes of their feasible ranges.

At the start of the problem solving process, there is usually very little information about the best settings of the experimental factors. Hopefully, a solution can be found within the feasible range. The strategy for moving the factor settings in the direction of the optimal settings is presented as part of the problem reduction strategies.

Experimental design and statistical analysis

We have already discussed selecting the best experimental design in Chapters 3 and 5 and conducting the statistical analysis in Chapters 4, 6, and 7. Application of these methods should lead to successful results for the design and analysis stages of the problem solving process.

8.4   Problem reduction

At the beginning of the process of solving an empirical problem we usually have many experimental factors and little information as to their optimal settings. The strategy that we use in going toward the goal of having few experimental factors and much information about their settings is called problem reduction.

Identifying important factors

At the initial stage of problem definition, we start with all potentially important experimental factors and our best guesses for their optimal settings. As many of these factors are included in the first experiment as is feasible. After the first and subsequent experiments, decisions are made as to which factors are clearly important, which factors are clearly unimportant, and which factors can not be classified as yet.

A factor is identified as important if its signal is greater than the noise. For example, factors may be identified based on review of the Pareto charts, normal plots and active contrast plots. The analysis of the reduced model should confirm the statistical significance of the important factors. Once an important factor has been identified, we want to change its settings to improve process performance. These strategies are described later in this section.

Selecting best values of qualitative factors

Qualitative factors are usually treated differently from quantitative factors because their settings cannot be easily adjusted from one experiment to the next. For example, a qualitative factor called surface type may have two reasonable values, say, polyurethane and polystyrene.

If a qualitative factor has been identified as being important, we usually just set it at its best value for subsequent experiments. In determining the best value, important interactions with other factors should be considered. Thus, once the best value has been determined, we no longer need to include the qualitative factor in the experimental design.

If a qualitative factor has not been shown to be significant yet, we usually would just include it again, unchanged in the next experiment. If it seems clear that the qualitative factor is not important, then we can set it at its most convenient or economic level, and we no longer need to include it in the next experiment.

Eliminating unimportant factors

By eliminating or discarding an experimental factor, we mean that in subsequent experiments we will fix it at a specific value. Thereafter, it will no longer be considered as a variable in the experimental design.

A factor can be at a specific level for one of the following two reasons:

- the factor doesn't affect the response or
- its "best" value has been determined.

Picking the best value is usually reserved for qualitative factors as we generally desire to make continued improvements in process performance by adjusting quantitative factors. Hence a quantitative factor will usually be eliminated only if it doesn't have an effect on process performance.

Some rules of thumb for deciding when a quantitative factor should be set to a fixed level due to lack of significance are as follows:

- if in two successive experiments with different ranges, the factor is not shown to be important.

- if in a single experiment, the factor has a very small effect on process performance, and there is prior believe that the factor may not be important.
- if in a single experiment, the factor is highly significant (very small p-value), and there are other reasons for setting it at a particular value.

The safest rule is to consider the results from more than one experiment as a means of reducing the chances of erroneously eliminating an important factor. However, it is fairly common to include factors which may be important but which you really hope aren't or which should if possible be set to a certain value. In these latter cases, the results from one experiment should be significant.

When eliminating a factor which doesn't appear to be important, we run the risk of committing a Type II error. For variables which are thought to be important, it is wise to change their ranges in a subsequent experiment before eliminating them because the wrong levels may have been chosen.

Moving toward optimal settings

Think of the response as a continuously varying landscape and the experimental region as the window through which we can view this landscape. Unfortunately, the middle of the window is blurred, and we can only see the landscape at the corners. However, we can move the window.

The strategy tells us the best direction and distance to move the window of the experimental region in response to the results of previous experiments. If we haven't moved far enough yet, we continue moving in the same direction for the next experiment. If we have gone too far, we move back.

In optimization experiments, we get a chance to look through the window at many more points. However, this requires much larger experiments so we only want to look if we are sure that the optimum settings are actually going to be viewable from where the window is located. Using this analogy, different experimental strategies might change

- the width of the window,
- the direction the window is moved, or
- how far the window is moved.

There are a number of strategies which can be used in choosing the experimental regions or "windows" in a series of experiments. They usually depend on "best guesses" concerning the possible effect each experimental factor may have on the response.

The following types of information may affect your strategy:

- the best guess as to the general location of the optimal settings; namely, if the likely optimal settings are
  * in the middle of the experimental limits,

* close to the experimental limits, or
* at the extremes of the experimental limits,
- knowledge of specific likely values for the optimal settings based on previous experience,
- a practical reason why you would like to exclude or include a particular ingredient or treatment unless there is a strong reason to do otherwise,
- whether you are looking only for large effects on the response or you are trying to "fine-tune" the response,
- experimental evidence of interactions among factors, or
- whether or not the experimental limits could be extended if there were a strong reason.

We will provide some explicit strategies to deal with these situations.

8.5   Strategies for achieving optimal performance

There are a number of different strategies which can be employed to change the settings of the experimental factors in order to improve process performance. As mentioned above, the choice of a strategy depends on what is known about the process and the particular experimental factors.

In this section, we present some sample strategies which have worked well for the types of problems which we work with most often. The reader should be aware of a competing approach to this problem which is called "the method of steepest ascent." Good descriptions of the steepest ascent methods can be found in Box, Hunter and Hunter (1978) and in Box and Draper (1987).

The method of steepest ascent calls for a calculating a path along which process performance is predicted to improve. A few exploratory points are observed along this path. The best point found along this path then serves as the center of the next experiment.

In comparison to the method of steepest ascent, the strategies we propose are more heuristic and do not involve the collection of exploratory data points. Rather, we generally find it more economical to use an explicit strategy for choosing new levels for the important factors based on prior information and on the results of the latest experiment. Then the next small experiment is conducted without intermediate exploratory observations.

A simple initial strategy

In order to set up a framework for the strategies to determine the experimental region, we consider a simple example. Suppose that one of the following conditions holds:

- either no prior information about the experimental factor is available or
- the best guess for the optimal settings is close to the center of the feasible region.

In either of these cases, we recommend the following strategy for determining the experimental regions in a series of screening experiments (Note: the high and low settings giving the experimental region will be specified as percentages of the feasible range, i.e., highest possible = 100%, lowest possible = 0%):

	Experimental Region	
	Low	High
Experiment #1:	40%	60%

	Experimental Region	
	Low	High
Experiment #2:		
A. If factor effect from the first experiment is positive	50%	80%
B. If factor effect from the first experiment is negative	20%	50%

If the effects remain positive in the first two experiments, then for the third experiment consider

	Experimental Region	
	Low	High
Experiment #3	70%	90%.

We leave it to the reader to suggest limits for the third experiment for each of the four possible outcomes of the first two experiments.

This strategy assumes that we are doing Resolution III experiments and so have no information about two-factor interactions. If a higher resolution design can be used, then information about two-factor interactions might influence the choice of levels in subsequent experiments. For example, it is possible that increasing both pH and the ionic strength of a buffer might harm a reaction whereas increasing either one separately would help. In this case, it would be necessary to study the interaction plots and to review the relative sizes of the estimates of the main effects and two-factor interactions before selecting levels for pH and ionic strength in the next experiment.

Evaluation of the first strategy

The important characteristics of this strategy are that it considers the direction of the effect of the experimental factor and then moves the experimental region in that direction. If there were different starting assumptions, an alternative strategy would be employed. Information which affects the choice of strategy include

- where the best guess settings are within the feasible region,
- whether the response is likely to get better or worse close to the boundaries of the feasible region,
- how quickly the response changes as a function of the experimental factor, and
- previous information about likely combinations of good settings.

For example, if we think that the response will fall off quickly when a particular factor gets close to its feasible limits, we would only choose settings close to its feasible limits if there were very strong evidence in favor of doing so.

Other strategies for maximizing a response

We present some other simple strategies to use in setting and moving the experimental region.

Assumption: The optimal settings are probably close to (but not equal to) one of the extremes of the experimental limits. It is not known which extreme is more likely.

Strategy: Assuming the coefficient for this factor in the first experiment is positive:

Experiment #1		_Experiment #2_	
Low	High	Low	High
20%	80%	70%	90%

Assumption: The optimal settings are probably at one of the extremes.

Strategy: Assuming the coefficient for this factor in the first experiment is positive:

Experiment #1		_Experiment #2_	
Low	High	Low	High
0%	100%	80%	90%

Assumption: The experimental factor represents an ingredient which you would not like to use unless necessary.

Strategy: Assuming the coefficient for this factor in the first experiment is not significant:

Experiment #1		_Experiment #2_	
Low	High	Low	High
0%	50%	0%	10%

Summary of strategy considerations

Here are some rules of thumb for determining a strategy for changing the settings of the experimental factors:

- start close to the best guess or else in the middle,
- overlap each range with the immediately previous one,
- continue moving ranges until
  - * the sign of the coefficient changes, in which case, the maximum has been passed or
  - * satisfactory process performance has been achieved.

Strategies to be used for minimizing a variable are just the reverse of those presented for maximizing a variable; i.e.,

OBJECTIVE	_EFFECT_	_ACTION_
Maximize	positive	increase settings
Maximize	negative	decrease settings
Minimize	positive	decrease settings
Minimize	negative	increase settings

There are two additional important considerations in implementing a strategy. First, an important factor may not have a statistically significant effect if the wrong range is chosen. In particular, an important factor can be missed if the initial settings are either too far apart (the response falls off at the extremes) or too close together (not enough difference to see the effect). Second, interactions between factors can also affect the strategy. That is, a large inconsistent interaction may be more important in determining the strategy than a smaller main effect.

Location of the maximum response

A particular strategy is intended to help find the settings of the optimal response. It reflects two key elements; namely,

- ranges are overlapped in successive experiments, and
- changes in the direction (or sign) of each effect are noted.

If we were to draw a simple example of a smooth response as a function of a single factor, we would notice that changes in the slope of the curve tell where the maximum response is located (Figure 8.2). That is, if we move from left to right, the slope changes from positive to negative.

The effect of each experimental factor will show up in the statistical analysis as an estimate of the slope of the response curve. Thus, when the effect changes sign, we know we have found the approximate location of the best settings. Unfortunately, this is not a foolproof method because of possible variations in where we set our experimental ranges and where the optimum is in relationship to the limits of the feasible region.

We overlap the ranges of successive experiments to decrease the chance that we might miss the optimum response. Two overlapping regions also provide more information together about the intersection of the two regions.

8.6 Summary

We discussed a number of strategies to use in solving empirical problems. In particular, when attempting to solve an empirical problem, the problem definition focuses our efforts on solving the right problem. At the beginning, we usually have

- many experimental factors and
- little information about the optimal settings.

Our objective in using an experimental strategy for problem reduction is to

- identify important experimental factors,
- eliminate unimportant experimental factors, and

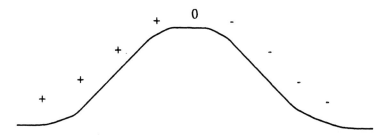

Figure 8.2 The slope of the response curve (coefficient of a factor effect) changes at the maximum response.

- get close to the optimal settings.

We use statistical experimental design and analysis

- to get the most information from the smallest experiments and
- to guide us in executing the experimental strategy.

In the context of a systematic problem solving strategy the tools for experimental design and statistical analysis can be very effective. In the next chapter, we tie together both strategy and statistical methods in a discussion of a real case study.

References and Bibliography

Box, G. E. P. and N. R. Draper (1987). *Empirical Model-Building and Response Surfaces.* New York: Wiley.

Box, G. E. P., W. G. Hunter, and J. S. Hunter (1978). *Statistics for Experimenters: An Introduction to Design, Data Analysis, and Model Building.* New York: Wiley.

Daniel, C. (1976). *Applications of Statistics to Industrial Experimentation.* New York: Wiley.

Joiner, B. L. (1981). Lurking variables: Some examples. *The American Statistician,* 35, 227-233.

Taguchi, G. (1986). *Introduction to Quality Engineering: Designing Quality Into Products and Processes.* Tokyo: Asian Productivity Organization.

Chapter 9

CASE HISTORY: STABILITY OF A MICROBEAD-BASED IMMUNOASSAY

Successful statistical problem solving is based on a series of small, carefully designed experiments. We are careful to ask sharp questions and to plan each step based on the facts revealed by the previous experiment. We use simple, yet powerful, tools based on statistical design and analysis. The unfolding nature of the problem solving process is presented in this case history.

9.1   Introduction

The iterative aspect of successful problem solving cannot be fully depicted by the study of single experiments. Thus, it is very difficult for a textbook to impart a feeling for the actual strategies used in statistical problem solving. However, in this case history we pay special attention to the strategy used in conducting a series of small experiments. We try to balance clarity of presentation with an accurate rendition of real life complexity so that the reader will find the case history useful and informative.

A series of experiments was used to solve a stability problem in a microbead-based immunoassay. Several responses and many experimental factors were studied. It was necessary to integrate the results from several experiments and to resolve conflicts in which one response could only be improved at the expense of degrading another. This complex problem was systematically investigated using statistical experimental design and analysis.

Description of problem

In this heterogeneous immunoassay, microbeads with antigen bound to their surface compete with antigen in the sample to bind with trapping antibody which has been bound to the surface. The signal is generated by fluorescent dye encapsulated within the microbeads. (See Figure 9.1.)

In this type of assay, the sample and microbeads are incubated together, and the greater the concentration of antigen in the sample, the fewer microbeads will be bound. Since the signal is proportional to the number of microbeads bound to the trapping antibodies, a high signal indicates no antigen in the sample and a low signal indicates the presence of antigen in the sample. Microbeads are used in this assay because unbound microbeads are easy to remove thereby eliminating any unbound fluorescent dye which would mask the presence of antigen in the sample.

A key issue in the stability of this immunoassay is the binding of the antigen to the microbeads. After a period of storage, it is possible for the antigen to come loose from the microbead. In this case, the microbeads would be washed away leaving no fluorescent dye to generate a signal even if there were no antigen in the sample. Thus a false positive would be generated. The

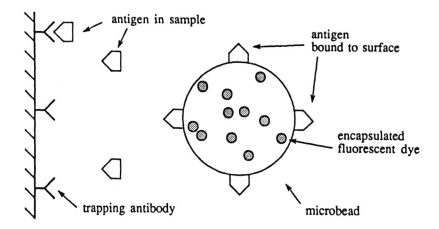

Figure 9.1 Schematic of a microbead with encapsulated fluorescent dye and with trapping antigen bound to its surface (not to scale).

medium in which the assay materials are stored plays an essential role in preserving this binding, and, hence, maintaining the stability of the assay materials.

In the early development stage of this assay, a storage buffer was identified which preserved assay performance. However, the concentration of microbeads in this storage buffer was too high to be used in the actual assay, so a more dilute buffer had to be developed. When the existing buffer was diluted to assay conditions, the assay materials degraded quickly. In fact, significant degradation could be observed at the end of one week of storage at room temperature in the dilute medium.

This stability problem had to be resolved before development could proceed further, so a series of experiments was conducted to solve the problem. The objective of these experiments was to identify a new formulation for the storage medium which would maintain the stability of the assay materials after dilution. The new storage medium must also be compatible with the assay materials in the sense of supporting good assay performance. It was not known *a priori* which of the constituents of the medium might affect the stability of the assay materials.

Stability criteria

In these experiments, all assay materials were initially stored in stock (undiluted) solution at 4°C. It was felt that only minimal degradation would occur at this temperature. An initial reading was taken in the stock solution (Time0), and then the solution was diluted to actual use conditions. After one week of storage at room temperature, a second reading was taken (Time1) and the percent of free antigen in the medium was determined (FreeAg). The difference between the initial and final readings (Delta) was also recorded.

The initial reading (Time0) was a zero binding reading; that is, the assay was conducted without introduction of a clinical sample. In this case, the microbeads should bind to all available antibody sites on the surface. The assay is configured so that if the stock solution is satisfactory, the zero binding reading should be about 0.050. If the antigen had separated from the microbeads in the storage medium, the signal-producing microbeads would be removed during one of the washes so that the zero binding reading would be less than 0.050.

If a dilute medium were a good storage medium, then the second reading, Time1, would also be close to 0.050. In which case, the difference between the readings, Delta, would be close to zero. Thus, the larger Delta is, the less stable the assay materials are in the dilute medium.

By taking measurements in both the stock and dilute media, it was hoped that general incompatibility problems could be separated from the stability of the assay materials. The measure of free antigen in the dilute medium after one week of storage in combination with the second zero binding reading should identify absolute separation of the microbeads from the antigen along with other effects on assay performance.

In order to judge the relative merits of the different storage media, the following criteria were applied to the response variables:

- Time0 should be maximized and ideally about 0.050.
- Time1 should be maximized and ideally about 0.050.
- Delta should be minimized and less than 0.010.
- FreeAg should be minimized and less than 1.0%.

There is no theoretical model which relates the performance of the assay on these criteria to the characteristics of the storage medium, so an empirical problem solving approach was employed.

Since there are several responses, it is possible that results for one of more of them may be conflicting; for example, a storage medium which preserved the antigen-to-microbead binding but which inhibited the antigen-to-antibody binding would minimize the free antigen (FreeAg), minimize the difference between the initial and follow-up zero binding readings (Time0 and Time1), but would also minimize both zero binding readings. Since it may not be possible to avoid such conflicts, it is necessary to carefully keep track of them and be prepared to make trade-offs among the response criteria if necessary.

Experimental factors

Very little was known about possible causes of the stability problem, and there were many potential experimental factors. An initial discussion among the research scientists and technicians identified twelve possible factors which might affect stability. However, some of these, based on prior knowledge, seem more likely than others to have important effects.

Upon reconsideration, eight of the twelve potential factors were thought to be important enough to include in a preliminary screening experiment. These eight factors are as follows:

- concentration of ascorbate (asc); strongly preferred value = 0.
- concentration of Ca^{++} (ca); prefer to include a small amount.
- concentration of phospholipase (phos); preferred value = 0, but a small amount may well be necessary.
- osmolality - similar to ionic strength but including all particles in solution (osm); standard value is 310, 380 has also been used in the past.
- concentration of DMSO (dmso); no preferred value.
- pH; should not be more acidic (lower) than pH 6.7 in order for the assay to work well and should not be more basic than pH 8.0.
- concentration of gelatin (gel); no preferred value.

- one of two types of buffer salts at 20 mM concentration (buf); Buffer Salt A is thought to work better than Buffer Salt B in the region of pH 6.7.

All of these factors are quantitative except for the type of buffer salt, so their ranges can be continuously adjusted. In the first experiment, each factor was set at two levels. For each factor, one level was at or near its preferred value.

9.2   An initial screening experiment

The usual goal of a screening experiment is problem reduction; that is, to separate the important factors from the unimportant factors and to suggest changes in factor settings which will improve process performance. In the first experiment, we also hope to identify one or two factors which seem to be much more important than the rest (recall the Pareto Principle which was presented in Chapter 1). Since we expect to need more than one experiment to solve this problem, we can make efficient use of available resources by starting off with a small experiment.

Experimental region

The initial experiment investigated eight factors, each at two-levels. Thus the experimental region is defined by a low value and a high value for each experimental factor. These values reflect prior information about each variable, are within the probable working limits of the medium, and avoid using zero and extreme values whenever possible. The values proposed for the first experiment are given in Table 9.1.

An explicit strategy was used in selecting the settings for each experimental factor; namely,

Table 9.1 Factor Levels for Initial Screening Experiment

Factor Name	Low (-1)	High (+1)
Ascorbate (asc)	0 mM	2 mM
Ca^{++} (ca)	1 mM	20 mM
Phospholipase (phos)	0.05%	0.5%
Osmolality (osm)	310	380
DMSO (dmso)	0.2%	2.0%
pH (pH)	6.7	8.0
Gelatin (gel)	0.1%	1.0%
Buffer salt type (buf)	A	B

- high and low levels should be far enough apart so that an effect, if it exists, is likely to be larger than the noise level.
- if there are two plausible limits or levels, they can be used as the high and low levels (e.g., pH, osmolality, and buffer type).
- if nothing is known about the potential effect, then subjectively choose values which represent "a little" and "a lot" (e.g., DMSO and gelatin).
- let one level be at or near the best guess or preferred value and let the other be noticeably higher or lower (e.g., Ca^{++}, and phospholipase).
- keep the high and low limits away from extreme values (e.g., "none" or "100%" at which the process may not work), unless there is a strong preference for the extreme value (e.g., ascorbate).

A Plackett-Burman design

Because this was a first experiment, it was desirable to use a small design which would efficiently separate the large signals (important factors) from the noise. Because of time constraints, only six different experimental media could be prepared per day although at least twelve runs could be assayed on a single day. Due to the cost of the assay materials and the extensive preparation time, it was desired to minimize the sample size.

The smallest designs in the Design Digest for eight factors require either 12 or 16 runs. A twelve run experiment could be prepared on two days whereas an eighteen run experiment would require three days to prepare. A twelve run experiment could also be assayed all on a single day.

There are no Resolution V designs for eight factors in the Design Digest (because such designs require more than 32 runs). The sixteen run design in eight factors (FF0816 in the Design Digest) is Resolution IV so there is confounding among two-factor interactions. The twelve run design in eight factors (PB0812) is Resolution III so there is also confounding among two-factor interactions and main effects. Because this is a first experiment with many factors and because of practical considerations, we chose to use the smaller, Resolution III design.

This twelve run design is called a Plackett-Burman design (Plackett and Burman, 1946). It allows us to investigate all of the factors in a single small experiment. Since this is a Resolution III design, we count on any important main effects being quite a bit larger than possible two-factor interactions. This is of some concern because of the potential interaction between pH and buffer (recall that buffer A is thought to be better at pH 6.7).

Because only half of the experimental media can be prepared on a single day, an extra factor, called day, was added to the experiment as a blocking variable. Thus, instead of PB0812, a variation of PB0912 was used. The worksheet for the Plackett-Burman design is given in Table 9.2.

Table 9.2 Worksheet for the Twelve Run, Eight Factor Plackett-Burman Design with Day as a Blocking Factor (PB0912)

Run	Asc	Ca	Phos	Osm	Dmso	pH	Gel	Buf	Day
1.	+1	+1	-1	-1	+1	-1	-1	-1	-1
2.	+1	-1	+1	-1	+1	+1	-1	-1	+1
3.	-1	+1	+1	-1	-1	+1	-1	+1	-1
4.	+1	+1	+1	+1	-1	+1	+1	-1	+1
5.	+1	+1	-1	+1	-1	-1	-1	+1	+1
6.	+1	-1	-1	+1	+1	+1	+1	+1	-1
7.	-1	-1	-1	-1	-1	-1	+1	-1	+1
8.	-1	-1	+1	+1	+1	-1	-1	+1	+1
9.	-1	+1	-1	-1	+1	+1	+1	+1	+1
10.	+1	-1	+1	-1	-1	-1	+1	+1	-1
11.	-1	+1	+1	+1	+1	-1	+1	-1	-1
12.	-1	-1	-1	+1	-1	+1	-1	-1	-1

Note: The factor day was used to divide this experiment into two blocks. Media preparations for runs with day=-1 were prepared on the first day while media for the day=+1 runs were prepared on the second day. The runs within each day were conducted in randomized order to guard against systematic bias.

In order to include some specific combinations of treatments in the design, PB0912 was not used exactly as shown in the Design Digest. Instead selected columns from PB1112 were used. In particular, we used columns A, K, J, -I, H, -G, F, E, respectively, for the experimental factors. Column D was used for the blocking factor (day). The minus signs indicate that we exchanged plus and minus ones from that column. The properties of a Plackett-Burman design are not affected by this type of substitution.

In order to check for two-factor interactions, we can use columns B and C of PB1112 as dummy variables in the statistical analysis. Since B and C do not correspond to any main effects, they estimate a combination of two-factor interactions (recall that this is a Resolution III design). However, if either of the B or C effects are large, we cannot identify which two-factor interactions are responsible because of the complicated confounding pattern of the Plackett-Burman design.

Results and analysis

The results from the first experiment are shown in Table 9.3. The most important result is readily apparent in that runs 1, 5, 7, 8, 10, and 11 had no free antigen (i.e., FreeAg=0). Looking back at Table 9.2 we see that these are all of

Table 9.3 Results from the First Experiment

Run	FreeAg	Time0	Time1	Delta
1.	0	.047	.032	.015
2.	11.7	.037	.009	.028
3.	11.7	.041	.014	.027
4.	8.4	.035	.011	.024
5.	0	.051	.029	.022
6.	14.6	.042	.012	.030
7.	0	.040	.042	-.002
8.	0	.054	.033	.021
9.	4.7	.038	.012	.026
10.	0	.047	.044	.003
11.	0	.041	.042	-.001
12.	15.9	.032	.014	.018

the runs at pH=6.7 (-1). The free antigen levels for all of the other runs are much higher than the 1% cutoff level given by our criteria. This is exactly the "big win" result we often get in a first experiment. This is an example of the factor sparsity aspect of the Pareto Principle.

Note that runs 7 and 11 had negative Delta values; that is Time1 was greater than Time0 for these two runs. We think that there was essentially no degradation for these samples and that these unusual values are the result of noise in the measurement process. We don't consider these values to be outliers since we have a plausible explanation for them.

We begin with the usual graphical analysis of the four responses; that is, Pareto charts (Figure 9.2), normal plots (Figure 9.3) and active contrast plots (Figure 9.4).

The lines on the normal plots were calculated by a method called least median of squared residuals (SPlus, 1988, and Rousseeuw and Leroy, 1987). This is a robust method for fitting a line through the middle part of the estimated effects (Lawson, 1988, and Nair, 1984). Effects which fall off the line, either below the line of the right hand side or above the line on the left hand side, are considered as possibly important effects.

In constructing the active contrast plots, we assume (based on our past experience) that as many as 40% of the effects may be important (active). We constructed active contrast plots for the following scale factors, 20, 20, 10, and 5, for each response. Our experience has been that the 40% prior probability of an active effect seems to fit the data well and emphasizes our concern with not wanting to miss any potentially important effects. We constructed plots for several scale factors and pick the one which seemed most informative, be-

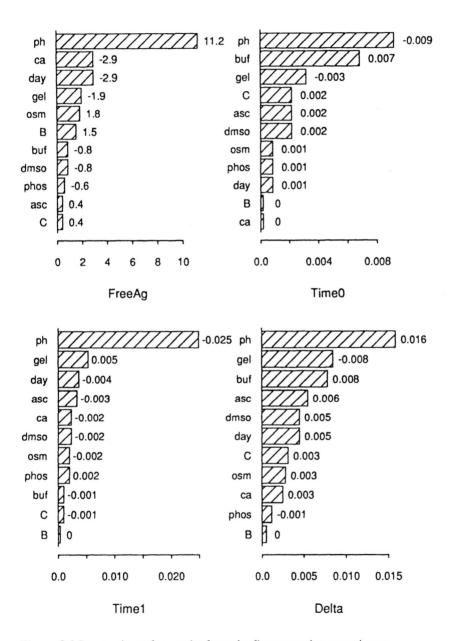

Figure 9.2 Pareto charts for results from the first screening experiment.

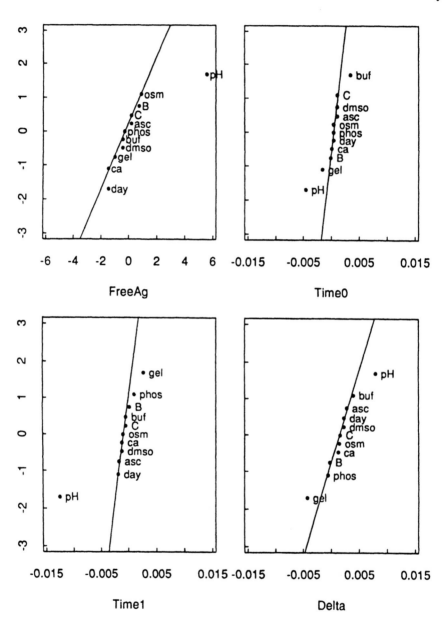

Figure 9.3 Normal plots for results from the first screening experiment.

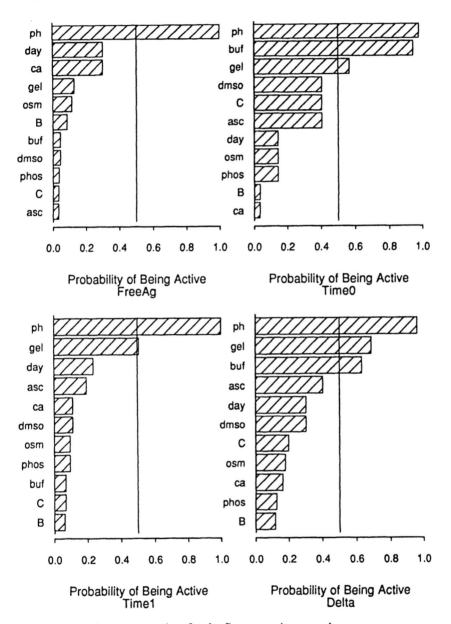

Figure 9.4 Active contrast plots for the first screening experiment.

cause there is often considerable variability from data set to data set with regard to the most informative choice of this parameter.

By far the most important factor in this study is pH. In particular, low pH completely eliminates dissociation of the antigen from the microbeads (i.e., FreeAg = 0 for each of the low pH runs). Low pH also maximizes Time0 and Time1 while minimizing Delta. Clearly, in the next experiment we should focus on low values of pH.

Among the other experimental factors, buffer salt type (buf) and gelatin concentration (gel) seem to be consistently important. Buffer salt type B maximizes Time0 (that is, the effect of buf is positive and the high level is type B). However, buffer salt type A minimizes Delta. Since it is suspected that the buffer salt type may depend on the pH, we should include both buffer salt types in a second experiment at lower pH levels in order to resolve this conflict.

A similar type of conflict is seen with the apparently important effect of gelatin (gel). In particular, using the higher level of gel increases the Time1 reading and decreases Delta while at the same time decreasing Time0. This suggests using higher levels of gelatin in the next experiment since the increase in Time1 offsets the decrease in Time0.

None of the other experimental factors seems to be large based on the graphical analysis. We are somewhat concerned that the blocking effect seems to be one of the larger effects for FreeAg, Time1, and Delta. It is not clear whether this is due to a true difference in preparation from day-to-day or to confounding with possible two-factor interactions. If there is a true day effect, the blocking has removed its influence from the estimates of the other effects.

The graphical analysis seems fairly clear at this point. We leave it to the reader to fit the reduced models. R-squared and t-values should be calculated and reviewed. The predicted versus observed plot and various residual plots should be examined to insure that the model fit is good, that there are no outliers, and that there is no indication that a transformation is required.

Conclusions

Of the initial eight factors, pH is clearly the most important. There is also good evidence that higher gelatin concentrations may be beneficial to assay stability. There was some indication that buffer type may also be important. At the levels used in this experiment, none of the remaining five factors seem to have an important effect on any of the responses.

Based on these results, two of these five other factors, Ca^{++} and osmolality, were fixed at convenient values in the next experiment for reasons other than their effect on assay performance. In particular, 310 is the standard value for osmolality for this assay buffer, and it now seems okay to continue its use. Also, it was desired to include a small amount of Ca^{++} in the storage medium because there are other reasons to believe that a small amount in the medium may contribute favorably to longer term stability.

At the concentrations used in the first experiment, the remaining three factors, ascorbate, phospholipase, and DMSO, were not important. In the next experiment, these three factors were considered at lower levels in the hope that they could be eliminated from the medium.

At this point, we reduced the number of factors to be included in the next experiment from eight to six. For the five quantitative factors still in consideration, values were selected for the second experiment which should either improve assay stability or be more economical. At this point no conclusions could be made about the best value of the qualitative factor (buffer salt type).

9.3   A second screening experiment

In the first experiment, we saw a good example of factor sparsity (Pareto Principle) which was discussed in Chapter 1. That is, a small initial experiment showed that only one out of the eight factors was mostly responsible for improving the stability of the assay storage medium. In the second experiment, adjustments were made in the most important factor, pH. The potential for the other five factors to make additional contributions to assay stability at the new settings of pH was also evaluated.

Experimental region

The second experiment also used a two-level, fractional factorial type design. Only 6 of the original 8 factors were included. The obvious choice for this design is the irregular fractional factorial IF0612. This is a nearly Resolution IV design which would allow estimation of a five two-factors interactions; for example, pH*gelatin.

However, at the time this experiment was done, we were not aware of this design. Consequently, a twelve run Plackett-Burman design was used. Due to the need for blocking to account for differences in media prepared on different days, a seventh blocking variable was again added to the design. Thus, design PB0712 was selected.

Table 9.4 shows the levels which were used in the second experiment. These choices were based on the conclusions reached from the analysis of the first experiment. For reasons of compatibility with other assay components, we wanted to stay above pH 6.7 so in the second experiment, we chose levels which were close to, but still higher than, pH 6.7. The values of gelatin were increased from the first experiment because of its positive effect. We decreased the values of ascorbate and phospholipase in order to see if they could be eliminated from the storage medium. A wider range was chosen for DM-

Table 9.4 Factor Levels for Second Screening Experiment

Factor Name	First Experiment		Second Experiment	
	Low (-1)	High (+1)	Low (-1)	High (+1)
Ascorbate (asc)	0 mM	2 mM	0 mM	1 mM
Ca^{++} (ca)	1 mM	20 mM	fixed at 1 mM	
Phospholipase (phos)	0.05%	0.5%	0%	0.1%
Osmolality (osm)	310	380	fixed at 310	
DMSO (dmso)	0.2%	2.0%	0%	0.1%
pH (ph)	6.7	8.0	7.0	7.5
Gelatin (gel)	0.1%	1.0%	0.7%	1.5%
Buffer salt type (buf)	A	B	A	B

SO. Buffer salt types remained the same. Notice in particular that calcium and osmolality are fixed at specific values.

The worksheet for PB0712 is shown in Table 9.5. The columns in Table 9.5 correspond to columns A, K, J, -I, H, G, and F, respectively, from PB1112. F is used as the blocking factor. The remaining columns from PB1112, namely, E,

Table 9.5 Worksheet for the Twelve Run, Six Factor Plackett-Burman Design with Day as a Blocking Factor (PB0712)

Run	Asc	Phos	Dmso	pH	Gel	Buf	Day
1.	+1	+1	-1	-1	+1	+1	-1
2.	+1	-1	+1	-1	+1	-1	-1
3.	-1	+1	+1	-1	-1	-1	-1
4.	+1	+1	+1	+1	-1	-1	+1
5.	+1	+1	-1	+1	-1	+1	-1
6.	+1	-1	-1	+1	+1	-1	+1
7.	-1	-1	-1	-1	-1	+1	+1
8.	-1	-1	+1	+1	+1	+1	-1
9.	-1	+1	-1	-1	+1	-1	+1
10.	+1	-1	+1	-1	-1	+1	+1
11.	-1	+1	+1	+1	+1	+1	+1
12.	-1	-1	-1	+1	-1	-1	-1

Note: The runs within each day were conducted in a randomized order.

D, B, and C will be used as dummy factors in the statistical analysis to serve as a check against unsuspected two-factor interactions.

Summary of results

The data collected in the second experiment are shown in Table 9.6. The usual Pareto charts, normal plots, and active contrast plots are shown in Figures 9.5, 9.6, and 9.7, respectively.

It appears that the important factors affecting FreeAg are buf, phos, D, ph, and asc. The important factors affecting Time0 are phos, gel, buf, and the dummy variable D. The important factors affecting Time1 are gel, ph, day, dmso and buf. The important factors affecting Delta are phos, buf, day, and ph.

Note that the lines on the normal plots (Figure 9.6) are not very informative for FreeAg, Time1, and Delta. We believe that this is because there are relatively more important factors than in many fractional factorial designs (compare to the first experiment in this series which is described in Section 9.2). A better alternative for drawing lines would be a simple linear regression through the middle 50% of the effects. This would work well for Time1 and Delta. It is probably not possible to draw an informative line on the normal plot for FreeAg.

A summary of the interpretation of this analysis is as follows:

*   ascorbate -- At the high level, this factor reduces the amount of free antigen in the medium after a week of storage (i.e., it has a negative co-efficient). It has no apparent effect on Time0, Time1 or Delta.
*   phospholipase -- The lower level of phospholipase (0%) resulted in an important reduction in free antigen. For Time0, the samples for

Table 9.6 Results from the Second Experiment

Run	FreeAg	Time0	Time1	Delta
1.	2.7	.038	.032	.006
2.	0	.046	.040	.006
3.	0	.026	.031	-.005
4.	0	.020	.016	.004
5.	0	.024	.015	.009
6.	2.9	.040	.021	.019
7.	2.7	.040	.019	.021
8.	6.1	.048	.030	.018
9.	0	.036	.028	.008
10.	4.2	.042	.019	.023
11.	6.1	.042	.022	.020
12.	4.6	.034	.017	.017

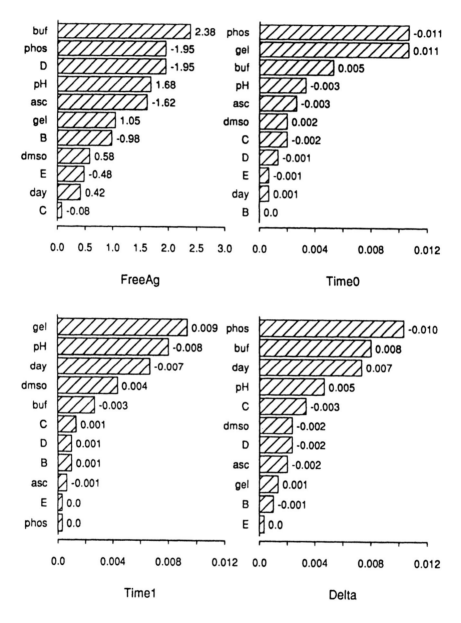

Figure 9.5 Pareto charts for results from the second screening experiment.

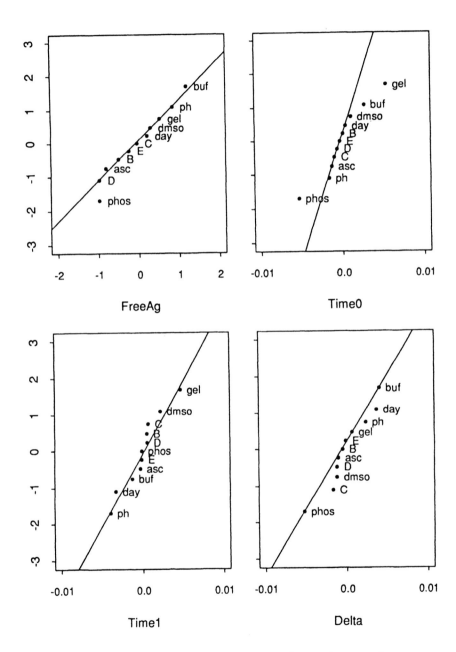

Figure 9.6 Normal plots for results from the second screening experiment.

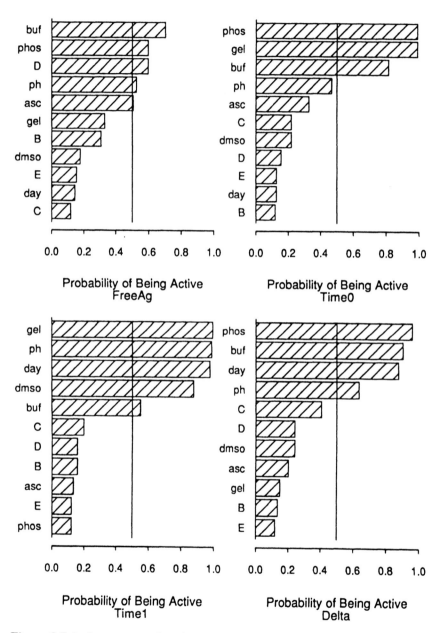

Figure 9.7 Active contrast plots for the second screening experiment.

phos=+1 have a higher mean value (.042) than those with phos=-1 (.031). However, this advantage is lost after a week of storage (Time1 mean value = .024 for phos=+1 and .024 for phos=-1).

- DMSO -- DMSO seems to have a slightly positive effect on Time1.
- pH -- Better results for FreeAg, Time1 and Delta are observed at pH 7 compared to pH 7.5.
- gelatin -- Gelatin is an important performance factor. However, it doesn't seem to affect FreeAg. The high level of gelatin had better average zero binding at both Time 0 and Time 1 (.042 and .029, respectively) than at the low level (.031 and .020, respectively). However, degradation was about the same at either level (averaging .013, gel=+1, versus .012, gel=-1).
- buffers -- Buffer salt type was the most important factor affecting free antigen, with type A being better. Buffer salt type A also seemed to improve stability. Although it had an adverse effect on initial performance (Time0 averaging .034 for buf=-1 versus .039 for buf=+1), after one week of storage the lower degradation associated with buffer salt type A resulted in higher average readings (Time1 averaging .026 for buf=-1 versus .023 for buf=+1).
- day -- As in the previous experiment, there appear to be day-to-day differences for Time1 and Delta. We cannot decide based on this design whether these are day effects or possible interactions.
- dummy variables -- The large effect of dummy variable D on free antigen suggests the possible presence of interactions. Such interactions would have to be investigated in a subsequent experiment using a higher resolution design.

Because of the large effects of day and the dummy variable D, we should examine the data carefully for the presence of one or more outliers. For example, run #3 is a suspected outlier due to its negative reading for Delta. This value is larger than either of the two negative values for Delta observed in the first experiment. However, an examination of predicted versus observed plots and residuals versus predicted plots for the reduced models (which we leave to the reader) doesn't show any evidence of outliers or unusual patterns in the data. Hence, as in the previous experiment, we attribute the result of run #3 to experimental error and leave it in the analysis. None of the other data values in Table 9.6 appear to be exceptional.

Conclusions

Combining what we have learned from these two experiments suggests that satisfactory assay performance and stability can be achieved at a pH level higher than 6.7 if other buffer characteristics are set appropriately (Table 9.7).

Table 9.7 Factor Levels for Second Screening Experiment

Factor Name	First Experiment		Second Experiment		"Best"
	Low	High	Low	High	
Ascorbate	0 mM	2 mM	0 mM	1 mM	1 mM
Ca^{++}	1 mM	20 mM	fixed at 1 mM		1 mM
Phospholipase	0.05%	0.5%	0%	0.1%	0%
Osmolality	310	380	fixed at 310		310
DMSO	0.2%	2.0%	0%	0.1%	0.1%
pH	6.7	8.0	7.0	7.5	7.0
Gelatin	0.1%	1.0%	0.7%	1.5%	1.5%
Buffer salt type	A	B	A	B	A

Given these results, it was felt that sufficient progress had been made so that no additional experiments were called for at this stage of product development.

If additional improvements in assay performance and stability were needed, it would be interesting to examine the factors DMSO concentration, pH and gelatin concentration in a response surface method or optimization experiment. It would make sense to center a central composite design (CC0318) around the best settings identified so far. We omit ascorbate from this list because given the current settings of the other factors, it seems to primarily affect free antigen levels, but satisfactory stability has already been achieved for this response.

9.4 Review of strategies

In the process of designing and carrying out these two experiments, we identified the important factors affecting assay performance and stability and we determined how to change the settings of these important factors in order to improve assay performance. If additional improvements in process performance were needed, we could proceed directly to an optimization experiment.

We also believe that we identified good values for the five factors no longer under investigation, and that we have a good idea of where the optimal settings for the remaining three factors are. Thus, efforts at problem reduction have been successful. We summarize the strategies used for problem reduction in the rest of this section.

Fixing factors at a particular value

When deciding to fix a factor at a particular value, we usually apply one of the following criteria:

- If the factor has no significant effect on the responses, then it can be set at a convenient, preferred or economical value.
- If a qualitative factor has a significant effect on the responses and does not interact with other factors, then fix the factor at its best value.

The first of these two strategies was used to select values for Ca^{++}, osmolality, and DMSO. For osmolality we wanted to use the standard value unless there was reason to do otherwise. For Ca^{++} we wanted to use a small amount for reasons of long term stability unless its use was contraindicated based on short term stability. DMSO had no effect on stability at all, so we set it at its economical level (zero).

Some caution should be shown in applying this rule since the high and low settings for an apparently unimportant variable may have been either too far apart, too close together, too high, or too low to result in an significant effect. Thus, it is often wise to include factors which are really thought to be important in a second experiment even if they are not significant in the first experiment. However, one experiment is usually sufficient to fix a value for a factor such as osmolality which was included in the first experiment more as a safeguard than as a potentially important factor.

The second of these criteria usually applies to qualitative factors such as buffer salt type for which there were only two types being considered and one of them was superior. We also apply this criteria to quantitative factors such as ascorbate when a limiting value (for example, zero) appears to be the best. If the qualitative factor does interact with other factors, then its best setting may change depending on the settings of other factors. In this case, the best settings of all of the interacting factors need to be evaluated jointly.

Changing factor settings in subsequent experiments

Each time we conduct another experiment, we generally change the low and high settings for each experimental factor. If we are trying to increase the response, we usually apply one of the following strategies:

- If the factor's coefficient is positive, we choose higher levels.
- If the factor coefficient is negative, we choose lower levels.
- If the factor coefficient is very close to zero, we might choose higher, lower, or wider levels.

In a series of experiments, we might apply one or more of these rules several times. Opposite strategies would be used if we were trying to minimize the response. For example, if a factor's coefficient were positive, we would choose lower levels to minimize the response.

For example, for phospholipase, we decreased the levels after the first and second experiment and concluded that it should not be included in the storage medium at all. When applying these rules to ascorbate, we kept the lower value fixed at its limit of zero (which was a preferred value). In the next experiment, we decreased the upper limit. The result of this change suggested that some ascorbate should be included in the medium.

For DMSO, we started at limits which represented "some" and "a lot." These limits were hopefully far enough apart to show an effect on stability if there were one. Since we didn't see any effect in the first experiment, we chose limits of "none" and "a little" for the second experiment. It turned out that "a little" DMSO seemed to help stability.

In our first experiment, we thought that pH might have a strong effect, and we chose limits which were quite far apart. We thought that the lower limit of pH 6.7 might be best, but we also had reason to hope that we could get a good result without going to such a low pH. The first experiment showed that pH 6.7 was much better than pH 8.0. Instead of bracketing pH 6.7 in the next experiment, we chose a fairly narrow region which was slightly above the lower limit. The results of the second experiment convinced us that a suitable storage medium could be developed at pH 7.0. However, additional improvements are probably still possible between pH 6.7 and pH 7.0.

Conclusion

The strategies we have applied in this chapter are very powerful. We applied them iteratively in using a series of experiments to solve a complex problem. These strategies are all also fairly simple and quite adaptable to a wide range of conditions and problems. They are especially well suited to iterative, systematic approaches to solving empirical problems.

References and Bibliography

Box, G. E. P., W. G. Hunter, and J. S. Hunter (1978). *Statistics for Experimenters: An Introduction to Design, Data Analysis, and Model Building.* New York: Wiley.

Daniel, C. (1976). *Applications of Statistics to Industrial Experimentation.* New York: Wiley.

Lawson, J. and L. Gold (1988). Robust estimation techniques for use in analysis of unreplicated 2^k and 2^{k-p} designs. Presented at the 1988 Annual Meetings of the American Statistical Association, New Orleans, LA.

Nair, V. N. (1984). On the behavior of some estimators from probability plots. *Journal of the American Statistical Association,* **79,** 823-831.

Plackett, R. L. and J. P. Burman (1946). The design of optimum multifactorial experiments. *Biometrika*, **33**, 305-325 and 328-332.

Rousseeuw, P. J. and A. M. Leroy (1987). *Robust Regression and Outlier Detection*. Wiley: New York.

SPlus (1988). Statistical Sciences, Inc., P. O. Box 85625, Seattle, WA 98145-1625.

Appendix

DESIGN DIGEST

1. Introduction and definitions

The purpose of the Design Digest is to assist the experimenter in selecting an experimental design. The core of this digest is the tables of designs for screening experiments and response surface experiments. The tables include information about design properties as well as listings of the designs. More detailed information on some design issues is included in the discussion sections.

Definitions

In order to clarify the description and discussion of design properties, we provide the following brief definitions of key terms:

Response variable: A measure of process performance, dependent variable.

Experimental factor: A process variable which the experimenter controls and which may affect process performance, an independent variable.

Run: After each of the process variables is set at a level of interest, the process is run and process performance is measured.

Experimental design: A collection of settings of the experimental factors at which the process will be run and response(s) measured.

189

Number of factors: The number of factors which are varied in the experimental design. This number does not include response variables.

Sample size: The number of runs called for by the experimental design. Each run yields exactly one value for the response variable(s)

Main effect: The difference in average process performance between the high and low settings of an experimental factor.

Two-factor interaction: Two factors are said to interact when the combined effect of the factors cannot be explained by the sum of their individual effects.

Two-factor interaction effect: The difference between the average of the response values when the two factors are at their extreme levels (high-high and low-low) and when the two factors are at mixed levels (high-low and low-high).

Quadratic effect: The effect due to the curvature term for a single experimental factor. An important quadratic effect indicates that the process performance is much better (or worse) at a midpoint between the high and low settings of one or more of the factors.

Confounded: Two effects are confounded when they cannot be estimated independently.

Confounding pattern: A typical confounding pattern is A*B+C*D (see design FF0408). This means that the two factor interaction A*B is confounded with the two factor interaction C*D. A*B and C*D cannot be distinguished from each other using this design. Only the sum of the two interactions can be estimated. If C*D is assumed to be negligible, then the estimate can be interpreted as the A*B interaction.

Resolution V: All main effects and all two-factor interactions can be estimated. No two-factor interactions are confounded with other two-factor interactions or main effects. No main effects are confounded with each other.

Resolution IV: No main effects are confounded with two-factor interactions, (or other main effects). Two-factor interactions may be confounded with other two-factor interactions.

Resolution III: No main effects are confounded with other main effects. Main effects may be confounded with two-factor interactions.

Randomization: Letting the sequence of trials be determined by chance to minimize the possibility of systematic errors due to uncontrolled variables that may change with time.

2. Quick reference tables for screening experiments

Sample sizes for standard fractional factorials

Table 1 lists the sample size required for a standard, two-level fractional factorial design of a specified resolution and number of factors.

Table 1. Sample Sizes for Standard Fractional Factorial Designs

Number of Factors	Sample Sizes Required for the Standard Two-Level Fractional Factorials		
	Resolution V	Resolution IV	Resolution III
3	8	*	4
4	16	8	*
5	16	*	8
6	32	16	8
7	64^+	16	8
8	64^+	16	*
9	128^+	32^+	16
10	128^+	32^+	16
11	128^+	32^+	16

* There is no standard two level fractional factorial design for this combination of resolution and number of factors.

+ This design is not included in the Design Digest.

NOTE: Smaller a sample size may sometimes be achieved using irregular fractions or Plackett-Burman designs (see Table 2).

Sample sizes for designs of fixed resolution

Table 2 lists the minimum sample size required for two-level screening designs in this digest having a specified resolution and number of factors

Table 2. Minimum Sample Sizes for Two-Level Designs with Specified Resolutions and Numbers of Factors

Number of Factors	Minimum Sample Sizes Required		
	Resolution V	Resolution IV	Resolution III
3	8	*	4
4	12^+	8	*
5	16	12^+	8
6	24^+	12^+	8
7	*	16	8
8	*	16	12
9	*	24	12
10	*	24	12
11	*	24	12

*   There is no design in the digest for this combination of resolution and number of factors.

\+   This design is *nearly* of the indicated resolution. See the discussion on irregular fractional factorials and resolution in Section 5.

Overview of designs in the digest

Table 3 provides a complete listing of all designs in the Design Digest.

Table 3. Listing of Experimental Designs in the Design Digest

	Design Name				
Number of Factors	8 Run Designs	12 Run Designs	16 Run Designs	24 Run Designs	32 Run Designs
3	FF0308	MF0312	*	*	*
4	FF0408 MF0412	IF0412 /or	FF0416	MF0424	*
5	FF0508 MF0512	IF0512/or	FF0516	MF0524	FF0532
6	FF0608	IF0612	FF0616	IF0624	FF0632
7	FF0708	PB0712	FF0716	PB0724	FF0732
8	+	PB0812	FF0816	PB0824	#
9	+	PB0912	FF0916	PB0924	#
10	+	PB1012	FF1016	PB1024	#
11	+	PB1112	FF1116	PB1124	#

* This design is not in the Design Digest, consider replicating a smaller design.
# This design is not in the Design Digest, consider using a 24 run design or contact a statistician for assistance
+ No design exists for this combination of factors and runs.

KEY: Design Name = aakkrr where
 aa=FF (two-level fractional factorial), MF (mixed two- and three-level fractional factorial), IF (irregular two-level fractional factorial), PB (Plackett-Burman)
 kk=number of factors, kk=3, 4, 5, ..., 11
 rr=number of runs, rr=8, 12, 16, 24, 32.

Number of two-factor interactions which can be estimated

Table 4 lists the maximum number of two-factor interactions that can be estimated for a specified number of factors and number of runs. See the specific design listings for further information.

Table 4. Number of Two-Factor Interactions Which Can Be Estimated

Number of Factors	Maximum Number of Interactions to be Estimated				
	8 Run Designs	12 Run Designs	16 Run Designs	24 Run Designs	32 Run Designs
3	3	3	*	*	*
4	3	6	6	6	*
5	2	5	10	10	10
6	1	4	7	15	15
7	0	0	7	6	18
8	+	0	7	7	#
9	+	0	6	8	#
10	+	0	5	9	#
11	+	0	4	10	#

*   This design is not in the digest, consider replicating a smaller design.
\#   This design is not in the digest, consider using a 24 run design or contact a statistician for assistance.
\+   No design exists for this combination of factors and runs.

3.   Listings of designs for screening experiments

This section includes the standard fractional factorial designs of 8, 16, and 32 runs and a variety of 12 and 24 run designs. These designs are especially useful for screening experiments. Screening experiments are typically used to determine which (out of possibly many) experimental factors affect process performance and how changing the settings of the important factors can improve process performance.

The designs are presented in the following order:

- Number of factors: from 3 to 11.
- Sample size: for each number of factors, the designs are ordered from smallest sample size to largest.

This digest obviously includes only a partial listing of designs. For information on selecting a design, see Chapters 3 and 5. The references from Chapters 1, 3, and 5 provide information about designs not included here.

Supplementary discussion of several interesting aspects of screening designs can be found in Sections 4-7. Section 4 discusses estimating effects in effect saturated designs. Section 5 discusses the resolutions of irregular and mixed-level fractional factorial designs. Center points and tests for curvature in screening experiments are discussed in Sections 6 and 7.

Number of Factors: 3 ....Design Name: FF0308

Resolution:	V		Factorial:	2^3
Sample Size:			Fraction:	Full
Base Design:	8			
Center Points:	3*			

Factors

run	A	B	C
1	-1	-1	-1
2	-1	-1	+1
3	-1	+1	-1
4	-1	+1	+1
5	+1	-1	-1
6	+1	-1	+1
7	+1	+1	-1
8	+1	+1	+1
9	0	0	0
10	0	0	0
11	0	0	0

Confounding Pattern: No confounding. All main effects and two-factor interactions can be estimated.

*   Include center points if possible (see Section 6).

Number of Factors: 3 .... **Design Name: MF0312**

Resolution:	V	Factorial:	$3*2^2$
Sample Size:	12	Fraction:	Full

Factors

run	A	B	C
1	-1	-1	-1
2	-1	-1	+1
3	-1	+1	-1
4	-1	+1	+1
5	0	-1	-1
6	0	-1	+1
7	0	+1	-1
8	0	+1	+1
9	+1	-1	-1
10	+1	-1	+1
11	+1	+1	-1
12	+1	+1	+1

Confounding Pattern: No confounding - all main effects, all two-factor interactions and 1 quadratic effect (A) can be estimated. Factor A can be quantitative or qualitative.

Number of Factors: 4.......................................Design Name: FF0408

| Resolution: | IV | | Factorial: | 2^4 |
| Sample Size: | | | Fraction: | 1/2 |

| Base Design: | 8 |
| Center Points: | 3* |

Factors

run	A	B	C	D
1	-1	-1	-1	-1
2	-1	-1	+1	+1
3	-1	+1	-1	+1
4	-1	+1	+1	-1
5	+1	-1	-1	+1
6	+1	-1	+1	-1
7	+1	+1	-1	-1
8	+1	+1	+1	+1
9	0	0	0	0
10	0	0	0	0
11	0	0	0	0

Confounding Pattern: Main effects (A to D) are not confounded with two-factor interactions. Two-factor interactions are confounded with other two-factor interactions. Only one two-factor interaction from each line listed in the confounding pattern given below can be estimated. (See Section 4.)

A*B + C*D
A*C + B*D
B*C + A*D

*   Include center points if possible (see Section 6).

Number of Factors: 4 ....Design Name: IF0412

Resolution:	nearly V*	Factorial:	2^4
Sample Size:	12	Fraction:	3/4

Factors

run	A	B	C	D
1	-1	-1	-1	-1
2	-1	-1	-1	+1
3	-1	-1	+1	-1
4	-1	-1	+1	+1
5	-1	+1	-1	-1
6	-1	+1	-1	+1
7	-1	+1	+1	-1
8	-1	+1	+1	+1
9	+1	-1	-1	+1
10	+1	-1	+1	-1
11	+1	+1	-1	-1
12	+1	+1	+1	+1

Confounding Pattern: All main effects and two-factor interactions can be estimated. There is some partial confounding, but it is limited and therefore is
not listed.

* See the discussion of the resolution of irregular fractional factorials in Section 5.

Number of Factors: 4 .... Design Name: MF0412

| Resolution: | nearly V* | | Factorial: | $3*2^3$ |
| Sample Size: | 12 | | Fraction: | 1/2 |

Factors

run	A	B	C	D
1	-1	-1	-1	-1
2	-1	-1	+1	+1
3	-1	+1	-1	+1
4	-1	+1	+1	-1
5	0	-1	-1	+1
6	0	-1	+1	-1
7	0	+1	-1	-1
8	0	+1	+1	+1
9	+1	-1	-1	-1
10	+1	-1	+1	+1
11	+1	+1	-1	+1
12	+1	+1	+1	-1

Confounding Pattern: All main effects, all two-factor interactions and 1 quadratic effect can be estimated. (See Section 4.) Some partial confounding
exists, but it is limited and therefore not listed. The following confounding
pattern applies for interactions with the quadratic effect (curvature term):

B*C + A²*D
B*D + A²*C
C*D + A²*B

* See the discussion of the resolution of mixed-level fractional factorials in
  Section 5.

Number of Factors: 4 .... Design Name: FF0416

Resolution:	V	Factorial:	2^4
Sample Size:		Fraction:	Full
Base Design:	16		
Center Points:	3*		

Factors

run	A	B	C	D
1	-1	-1	-1	-1
2	-1	-1	-1	+1
3	-1	-1	+1	-1
4	-1	-1	+1	+1
5	-1	+1	-1	-1
6	-1	+1	-1	+1
7	-1	+1	+1	-1
8	-1	+1	+1	+1
9	+1	-1	-1	-1
10	+1	-1	-1	+1
11	+1	-1	+1	-1
12	+1	-1	+1	+1
13	+1	+1	-1	-1
14	+1	+1	-1	+1
15	+1	+1	+1	-1
16	+1	+1	+1	+1
17	0	0	0	0
18	0	0	0	0
19	0	0	0	0

Confounding Pattern: No confounding. All main effects and two-factor interactions can be estimated.

* Include center points if possible (see Section 6).

Number of Factors: 4 .... **Design Name: MF0424**

Resolution:	V		Factorial:	$3*2^3$
Sample Size			Fraction:	Full
Base Design:	24			
Center Points:	3*			

Factors

run	A	B	C	D
1	-1	-1	-1	-1
2	-1	-1	-1	+1
3	-1	-1	+1	-1
4	-1	-1	+1	+1
5	-1	+1	-1	-1
6	-1	+1	-1	+1
7	-1	+1	+1	-1
8	-1	+1	+1	+1
9	0	-1	-1	-1
10	0	-1	-1	+1
11	0	-1	+1	-1
12	0	-1	+1	+1
13	0	+1	-1	-1
14	0	+1	-1	+1
15	0	+1	+1	-1
16	0	+1	+1	+1
17	+1	-1	-1	-1
18	+1	-1	-1	+1
19	+1	-1	+1	-1
20	+1	-1	+1	+1
21	+1	+1	-1	-1
22	+1	+1	-1	+1
23	+1	+1	+1	-1
24	+1	+1	+1	+1
25	0	0	0	0
26	0	0	0	0
27	0	0	0	0

Confounding Pattern: No confounding. All main effects, all two-factor interactions and 1 quadratic effect (of A) can be estimated. All interactions between the quadratic effect and other main effects can be estimated (for example, A^2*B, etc.).

* Include center points if possible (see Section 6).

Number of Factors: 5 ....Design Name: FF0508

| Resolution: | III | | Factorial: | 2^5 |
| Sample Size: | 8 | | Fraction: | 1/4 |

Factors

run	A	B	C	D	E
1	-1	-1	-1	-1	+1
2	-1	-1	+1	+1	-1
3	-1	+1	-1	+1	-1
4	-1	+1	+1	-1	+1
5	+1	-1	-1	+1	+1
6	+1	-1	+1	-1	-1
7	+1	+1	-1	-1	-1
8	+1	+1	+1	+1	+1

Confounding Pattern: Main effects are confounded with two-factor interactions. Only one effect from each line listed in the confounding pattern given below can be estimated. The maximum number of effects that can be estimated is 7 (5 main effects and 2 interactions). (See Section 4.)

A + D*E
B + C*E
C + B*E
D + A*E
E + B*C + A*D
   A*B + C*D
   A*C + B*D

Number of Factors: 5................................... **Design Name: MF0512**

Resolution:	nearly IV*	Factorial: $3*2^4$
Sample Size:	12	Fraction: 1/4

Factors

run	A	B	C	D	E
1	-1	-1	-1	-1	-1
2	-1	-1	+1	+1	-1
3	-1	+1	-1	-1	+1
4	-1	+1	+1	+1	+1
5	0	-1	-1	+1	+1
6	0	-1	+1	-1	+1
7	0	+1	-1	+1	-1
8	0	+1	+1	-1	-1
9	+1	-1	-1	-1	-1
10	+1	-1	+1	+1	-1
11	+1	+1	-1	-1	+1
12	+1	+1	+1	+1	+1

Confounding Pattern: There is some slight confounding among main effects. Main effects are completely unconfounded with two-factor interactions. Only one effect from each line below can be estimated. A maximum of 10 effects can be estimated (5 main effects, 1 quadratic effect and 4 interactions). Interactions with the quadratic effect are confounded with main effects and can not be estimated.

$A^2 + C*D + B*E$
   $A*B + A*E$
   $A*C + A*D$
   $B*C + D*E$
   $B*D + C*E$

* See the discussion of mixed-level fractional factorials and resolution in Section 5.

Number of Factors: 5...................................Design Name: IF0512

	Resolution:	nearly IV*		Factorial:	2^5
	Sample Size:	12		Fraction:	3/8

Factors

run	A	B	C	D	E
1	-1	-1	-1	+1	+1
2	-1	-1	+1	-1	-1
3	-1	+1	-1	-1	+1
4	-1	+1	-1	+1	-1
5	-1	+1	+1	-1	+1
6	-1	+1	+1	+1	-1
7	+1	-1	-1	-1	+1
8	+1	-1	-1	+1	-1
9	+1	-1	+1	-1	+1
10	+1	-1	+1	+1	-1
11	+1	+1	-1	+1	+1
12	+1	+1	+1	-1	-1

Confounding Pattern: There is some slight confounding among main effects. Main effects are completely unconfounded with two-factor interactions. Only one two-factor interaction from each line below can be estimated. A total of two interactions from group 1 and three interactions from group 2 can be estimated.

A*B + D*E
C*D   - group 1
C*E

A*C
B*C   - group 2
A*D + B*E
B*D + A*E

* See the discussion of the resolution of irregular fractional factorials in Section 5.

Number of Factors: 5 .... Design Name: FF0516

Resolution:	V	Factorial:	2^5
Sample Size:		Fraction:	1/2
Base Design:	16		
Center Points:	3*		

Factors

run	A	B	C	D	E
1	-1	-1	-1	-1	+1
2	-1	-1	-1	+1	-1
3	-1	-1	+1	-1	-1
4	-1	-1	+1	+1	+1
5	-1	+1	-1	-1	-1
6	-1	+1	-1	+1	+1
7	-1	+1	+1	-1	+1
8	-1	+1	+1	+1	-1
9	+1	-1	-1	-1	-1
10	+1	-1	-1	+1	+1
11	+1	-1	+1	-1	+1
12	+1	-1	+1	+1	-1
13	+1	+1	-1	-1	+1
14	+1	+1	-1	+1	-1
15	+1	+1	+1	-1	-1
16	+1	+1	+1	+1	+1
17	0	0	0	0	0
18	0	0	0	0	0
19	0	0	0	0	0

Confounding Pattern: No confounding. All main effects and two-factor interactions can be estimated.

* Include center points if possible (see Section 6).

Number of Factors: 5 .... **Design Name: MF0524**

Resolution:	nearly V*	Factorial:	$3*2^4$
Sample Size:		Fraction:	1/2
Base Design:	24		
Center Points:	3^+		

Factors

run	A	B	C	D	E
1	-1	-1	-1	-1	-1
2	-1	-1	-1	+1	+1
3	-1	-1	+1	-1	+1
4	-1	-1	+1	+1	-1
5	-1	+1	-1	-1	+1
6	-1	+1	-1	+1	-1
7	-1	+1	+1	-1	-1
8	-1	+1	+1	+1	+1
9	0	-1	-1	-1	+1
10	0	-1	-1	+1	-1
11	0	-1	+1	-1	-1
12	0	-1	+1	+1	+1
13	0	+1	-1	-1	-1
14	0	+1	-1	+1	+1
15	0	+1	+1	-1	+1
16	0	+1	+1	+1	-1
17	+1	-1	-1	-1	-1
18	+1	-1	-1	+1	+1
19	+1	-1	+1	-1	+1
20	+1	-1	+1	+1	-1
21	+1	+1	-1	-1	+1
22	+1	+1	-1	+1	-1
23	+1	+1	+1	-1	-1
24	+1	+1	+1	+1	+1
25	0	0	0	0	0
26	0	0	0	0	0
27	0	0	0	0	0

Confounding Pattern: There is some limited partial confounding. All main effects, all two-factor interactions and 1 quadratic effect (of A) can be estimated. All interactions between the quadratic effect and other main effects can be estimated (for example, A^2*B, etc.)

+   Include center points if possible (see Section 6).
*   See the discussion on resolution of mixed-level designs in Section 5.

Number of Factors: 5.....................................Design Name: FF0532

	Resolution:	V		Factorial:	2^5
	Sample Size:			Fraction:	Full
	Base Design:	32			
	Center Points:	3*			

			Factors						Factors		
run	A	B	C	D	E	run	A	B	C	D	E
1	-1	-1	-1	-1	-1	19	+1	-1	-1	+1	-1
2	-1	-1	-1	-1	+1	20	+1	-1	-1	+1	+1
3	-1	-1	-1	+1	-1	21	+1	-1	+1	-1	-1
4	-1	-1	-1	+1	+1	22	+1	-1	+1	-1	+1
5	-1	-1	+1	-1	-1	23	+1	-1	+1	+1	-1
6	-1	-1	+1	-1	+1	24	+1	-1	+1	+1	+1
7	-1	-1	+1	+1	-1	25	+1	+1	-1	-1	-1
8	-1	-1	+1	+1	+1	26	+1	+1	-1	-1	+1
9	-1	+1	-1	-1	-1	27	+1	+1	-1	+1	-1
10	-1	+1	-1	-1	+1	28	+1	+1	-1	+1	+1
11	-1	+1	-1	+1	-1	29	+1	+1	+1	-1	-1
12	-1	+1	-1	+1	+1	30	+1	+1	+1	-1	+1
13	-1	+1	+1	-1	-1	31	+1	+1	+1	+1	-1
14	-1	+1	+1	-1	+1	32	+1	+1	+1	+1	+1
15	-1	+1	+1	+1	-1	33	0	0	0	0	0
16	-1	+1	+1	+1	+1	34	0	0	0	0	0
17	+1	-1	-1	-1	-1	35	0	0	0	0	0
18	+1	-1	-1	-1	+1						

Confounding Pattern: No confounding. All main effects and two-factor interactions can be estimated.

* Include center points if possible (see Section 6).

Number of Factors: 6 .... Design Name: FF0608

Resolution:	III	Factorial:	2^6
Sample Size:	8	Fraction:	1/8

Factors

run	A	B	C	D	E	F
1	-1	-1	-1	-1	+1	+1
2	-1	-1	+1	+1	-1	-1
3	-1	+1	-1	+1	-1	+1
4	-1	+1	+1	-1	+1	-1
5	+1	-1	-1	+1	+1	-1
6	+1	-1	+1	-1	-1	+1
7	+1	+1	-1	-1	-1	-1
8	+1	+1	+1	+1	+1	+1

Confounding Pattern: Main effects are confounded with two-factor interactions. Only one effect from each line below can be estimated. The maximum number of effects that can be estimated is 7 (6 main effects and 1 interaction).

A + D*E + C*F
B + C*E + D*F
C + B*E + A*F
D + A*E + B*F
E + B*C + A*D
F + A*C + B*D
    A*B + C*D + E*F

Number of Factors: 6.......................................**Design Name: IF0612**

| | Resolution: | nearly IV* | | | Factorial: | 2^6 |
| | Sample Size: | 12 | | | Fraction: | 3/16 |

Factors

run	A	B	C	D	E	F
1	-1	-1	-1	+1	+1	+1
2	-1	-1	+1	-1	-1	+1
3	-1	+1	-1	-1	+1	-1
4	-1	+1	-1	+1	-1	+1
5	-1	+1	+1	-1	+1	+1
6	-1	+1	+1	+1	-1	-1
7	+1	-1	-1	-1	+1	+1
8	+1	-1	-1	+1	-1	-1
9	+1	-1	+1	-1	+1	-1
10	+1	-1	+1	+1	-1	+1
11	+1	+1	-1	+1	+1	-1
12	+1	+1	+1	-1	-1	-1

Confounding Pattern: There is some slight confounding among main effects. Main effects are completely unconfounded with two-factor interactions. Only one two-factor interaction from each line below can be estimated. A total of two interactions from group 1 and three interactions from group 2 can be estimated. (See Section 4.)

A*B + D*E
C*D + A*F   - group 1
C*E + B*F

A*C + D*F
B*C + E*F   - group 2
A*D + B*E + C*F
B*D + A*E

* See the discussion of the resolution of irregular fractional factorials in Section 5.

Number of Factors: 6 ....**Design Name: FF0616**

Resolution: IV	Factorial:	2^6
Sample Size:	Fraction:	1/4
Base Design: 16		
Center Points: 3*		

Factors

run	A	B	C	D	E	F
1	-1	-1	-1	-1	-1	-1
2	-1	-1	-1	+1	+1	+1
3	-1	-1	+1	-1	+1	+1
4	-1	-1	+1	+1	-1	-1
5	-1	+1	-1	-1	+1	-1
6	-1	+1	-1	+1	-1	+1
7	-1	+1	+1	-1	-1	+1
8	-1	+1	+1	+1	+1	-1
9	+1	-1	-1	-1	-1	+1
10	+1	-1	-1	+1	+1	-1
11	+1	-1	+1	-1	+1	-1
12	+1	-1	+1	+1	-1	+1
13	+1	+1	-1	-1	+1	+1
14	+1	+1	-1	+1	-1	-1
15	+1	+1	+1	-1	-1	-1
16	+1	+1	+1	+1	+1	+1
17	0	0	0	0	0	0
18	0	0	0	0	0	0
19	0	0	0	0	0	0

Confounding Pattern: Main effects (A through F) are not confounded with two-factor interactions. Only one two-factor interaction from each line below can be estimated.

    A*B + E*F
    A*C + D*F
    B*C + D*E
    A*D + C*F
    B*D + C*E
    A*E + B*F
    C*D + B*E + A*F

* Include center points if possible (see Section 6).

Number of Factors: 6 ....**Design Name: IF0624**

Resolution:	nearly V*	Factorial: 2^6
Sample Size:		Fraction: 3/8
Base Design:	24	
Center Points:	3^+	

Factors

run	A	B	C	D	E	F
1	-1	-1	-1	-1	+1	+1
2	-1	-1	-1	+1	-1	-1
3	-1	-1	+1	-1	-1	-1
4	-1	-1	+1	+1	+1	+1
5	-1	+1	-1	-1	+1	-1
6	-1	+1	-1	+1	-1	+1
7	-1	+1	+1	-1	-1	+1
8	-1	+1	+1	+1	+1	-1
9	+1	-1	-1	-1	-1	+1
10	+1	-1	-1	+1	-1	-1
11	+1	-1	-1	-1	+1	-1
12	+1	-1	-1	+1	+1	+1
13	+1	+1	-1	-1	-1	-1
14	+1	+1	-1	+1	-1	+1
15	+1	+1	-1	-1	+1	+1
16	+1	+1	-1	+1	+1	-1
17	+1	-1	+1	-1	-1	-1
18	+1	-1	+1	+1	-1	+1
19	+1	-1	+1	-1	+1	+1
20	+1	-1	+1	+1	+1	-1
21	+1	+1	+1	-1	-1	+1
22	+1	+1	+1	+1	-1	-1
23	+1	+1	+1	-1	+1	-1
24	+1	+1	+1	+1	+1	+1
25	0	0	0	0	0	0
26	0	0	0	0	0	0
27	0	0	0	0	0	0

Confounding Pattern: All main effects and two-factor interactions can be estimated. There is some partial confounding, but it is limited and therefore not
listed. If minimizing sample size is not critical, a 32 run design is recommended.

* See the discussion of the resolution of irregular fractional factorials in Section 5.

+ Include center points if possible (see Section 6).

Number of Factors: 6.....................................**Design Name: FF0632**

Resolution:	V			Factorial:	2^6
Sample Size:				Fraction:	1/2
Base Design:	32				
Center Points:	3*				

Factors

run	A	B	C	D	E	F
1	-1	-1	-1	-1	-1	-1
2	-1	-1	-1	-1	+1	+1
3	-1	-1	-1	+1	-1	+1
4	-1	-1	-1	+1	+1	-1
5	-1	-1	+1	-1	-1	+1
6	-1	-1	+1	-1	+1	-1
7	-1	-1	+1	+1	-1	-1
8	-1	-1	+1	+1	+1	+1
9	-1	+1	-1	-1	-1	+1
10	-1	+1	-1	-1	+1	-1
11	-1	+1	-1	+1	-1	-1
12	-1	+1	-1	+1	+1	+1
13	-1	+1	+1	-1	-1	-1
14	-1	+1	+1	-1	+1	+1
15	-1	+1	+1	+1	-1	+1
16	-1	+1	+1	+1	+1	-1
17	+1	-1	-1	-1	-1	+1
18	+1	-1	-1	-1	+1	-1
19	+1	-1	-1	+1	-1	-1
20	+1	-1	-1	+1	+1	+1
21	+1	-1	+1	-1	-1	-1
22	+1	-1	+1	-1	+1	+1
23	+1	-1	+1	+1	-1	+1
24	+1	-1	+1	+1	+1	-1

Number of Factors: 6......................................**Design Name: FF0632**

(continued from previous page.)

Factors

run	A	B	C	D	E	F
26	+1	+1	-1	-1	+1	+1
27	+1	+1	-1	+1	-1	+1
28	+1	+1	-1	+1	+1	-1
29	+1	+1	+1	-1	-1	+1
30	+1	+1	+1	-1	+1	-1
31	+1	+1	+1	+1	-1	-1
32	+1	+1	+1	+1	+1	+1
33	0	0	0	0	0	0
34	0	0	0	0	0	0
35	0	0	0	0	0	0

Confounding Pattern: No confounding. All main effects and two-factor interactions can be estimated.

* Include center points if possible (see Section 6).

Number of Factors: 7..**Design Name: FF0708**

| | Resolution: | III | | | Factorial: | | 2^7 |
| | Sample Size: | 8 | | | Fraction: | | 1/16 |

Factors

run	A	B	C	D	E	F	G
1	-1	-1	-1	-1	+1	+1	+1
2	-1	-1	+1	+1	-1	-1	+1
3	-1	+1	-1	+1	-1	+1	-1
4	-1	+1	+1	-1	+1	-1	-1
5	+1	-1	-1	+1	+1	-1	-1
6	+1	-1	+1	-1	-1	+1	-1
7	+1	+1	-1	-1	-1	-1	+1
8	+1	+1	+1	+1	+1	+1	+1

Confounding Pattern: Main effects are confounded with two-factor interactions. All 7 main effects can be estimated when no two-factor interactions are present. (See Section 4.) No two-factor interactions can be included in the model.

A + D*E + C*F + B*G
B + C*E + A*G + D*F
C + B*E + A*F + D*G
D + A*E + B*F + C*G
E + B*C + A*D + F*G
F + A*C + B*D + E*G
G + A*B + C*D + E*F

Number of Factors: 7 .... Design Name: PB0712

Resolution:	III	Factorial:	2^7
Sample Size:	12	Fraction:	3/32

Factors

run	A	B	C	D	E	F	G
1	+1	-1	+1	-1	-1	-1	+1
2	+1	+1	-1	+1	-1	-1	-1
3	-1	+1	+1	-1	+1	-1	-1
4	+1	-1	+1	+1	-1	+1	-1
5	+1	+1	-1	+1	+1	-1	+1
6	+1	+1	+1	-1	+1	+1	-1
7	-1	+1	+1	+1	-1	+1	+1
8	-1	-1	+1	+1	+1	-1	+1
9	-1	-1	-1	+1	+1	+1	-1
10	+1	-1	-1	-1	+1	+1	+1
11	-1	+1	-1	-1	-1	+1	+1
12	-1	-1	-1	-1	-1	-1	-1

Confounding Pattern: All main effects (A through G) are partially confounded with two-factor interactions. Main effects are not confounded with other main effects.

Number of Factors: 7.....................................**Design Name: FF0716**

Resolution:	IV		Factorial:	2^7
Sample Size:			Fraction:	1/8
Base Design:	16			
Center Points:	3*			

Factors

run	A	B	C	D	E	F	G
1	-1	-1	-1	-1	-1	-1	-1
2	-1	-1	-1	+1	+1	+1	+1
3	-1	-1	+1	-1	+1	+1	-1
4	-1	-1	+1	+1	-1	-1	+1
5	-1	+1	-1	-1	+1	-1	+1
6	-1	+1	-1	+1	-1	+1	-1
7	-1	+1	+1	-1	-1	+1	+1
8	-1	+1	+1	+1	+1	-1	-1
9	+1	-1	-1	-1	-1	+1	+1
10	+1	-1	-1	+1	+1	-1	-1
11	+1	-1	+1	-1	+1	-1	+1
12	+1	-1	+1	+1	-1	+1	-1
13	+1	+1	-1	-1	+1	+1	-1
14	+1	+1	-1	+1	-1	-1	+1
15	+1	+1	+1	-1	-1	-1	-1
16	+1	+1	+1	+1	+1	+1	+1
17	0	0	0	0	0	0	0
18	0	0	0	0	0	0	0
19	0	0	0	0	0	0	0

Confounding Pattern: Main effects (A through G) are not confounded with two-factor interactions. Only one two-factor interaction from each line below can be estimated.

A*B + E*F + D*G
A*C + D*F + E*G
B*C + D*E + F*G
A*D + C*F + B*G
B*D + C*E + A*G
A*E + B*F + C*G
C*D + B*E + A*F

*Include center points if possible (see Section 6).

Number of Factors: 7 .... **Design Name: PB0724$^+$**

Resolution:	IV	Factorial:	2^7
Sample Size:		Fraction:	3/16
Base Design:	24		
Center Points:	3*		

<u>Factors</u>

run	A	B	C	D	E	F	G
1	-1	+1	-1	+1	-1	-1	-1
2	-1	+1	+1	-1	+1	-1	-1
3	-1	-1	+1	+1	-1	+1	-1
4	-1	+1	-1	+1	+1	-1	+1
5	-1	+1	+1	-1	+1	+1	-1
6	-1	+1	+1	+1	-1	+1	+1
7	-1	-1	+1	+1	+1	-1	+1
8	-1	-1	-1	+1	+1	+1	-1
9	-1	-1	-1	-1	+1	+1	+1
10	-1	+1	-1	-1	-1	+1	+1
11	-1	-1	+1	-1	-1	-1	+1
12	-1	-1	-1	-1	-1	-1	-1
13	+1	-1	+1	-1	+1	+1	+1
14	+1	-1	-1	+1	-1	+1	+1
15	+1	+1	-1	-1	+1	-1	+1
16	+1	-1	+1	-1	-1	+1	-1
17	+1	-1	-1	+1	-1	-1	+1
18	+1	-1	-1	-1	+1	-1	-1
19	+1	+1	-1	-1	-1	+1	-1
20	+1	+1	+1	-1	-1	-1	+1
21	+1	+1	+1	+1	-1	-1	-1
22	+1	-1	+1	+1	+1	-1	-1
23	+1	+1	-1	+1	+1	+1	-1
24	+1	+1	+1	+1	+1	+1	+1
25	0	0	0	0	0	0	0
26	0	0	0	0	0	0	0
27	0	0	0	0	0	0	0

<u>Confounding Pattern</u>: Main effects are not confounded with two-factor interactions. All two-factor interactions are partially confounded with other two-factor interactions. All two-factor interactions with factor A can be estimated.

* Include center points if possible (see Section 6).

$^+$ This is a "fold-over" of a 12 run Plackett-Burman design. See the discussion in Section 5.

Number of Factors: 7 ....Design Name: FF0732

	Resolution:	IV			Factorial:		2^7
	Sample Size:				Fraction:		1/4
	Base Design:		32				
	Center Points:		3*				

<u>Factors</u>

run	A	B	C	D	E	F	G
1	-1	-1	-1	-1	-1	+1	+1
2	-1	-1	-1	-1	+1	-1	+1
3	-1	-1	-1	+1	-1	+1	-1
4	-1	-1	-1	+1	+1	-1	-1
5	-1	-1	+1	-1	-1	-1	-1
6	-1	-1	+1	-1	+1	+1	-1
7	-1	-1	+1	+1	-1	-1	+1
8	-1	-1	+1	+1	+1	+1	+1
9	-1	+1	-1	-1	-1	-1	-1
10	-1	+1	-1	-1	+1	+1	-1
11	-1	+1	-1	+1	-1	-1	+1
12	-1	+1	-1	+1	+1	+1	+1
13	-1	+1	+1	-1	-1	+1	+1
14	-1	+1	+1	-1	+1	-1	+1
15	-1	+1	+1	+1	-1	+1	-1
16	-1	+1	+1	+1	+1	-1	-1
17	+1	-1	-1	-1	-1	-1	-1
18	+1	-1	-1	-1	+1	+1	-1
19	+1	-1	-1	+1	-1	-1	+1
20	+1	-1	-1	+1	+1	+1	+1
21	+1	-1	+1	-1	-1	+1	+1
22	+1	-1	+1	-1	+1	-1	+1
23	+1	-1	+1	+1	-1	+1	-1
24	+1	-1	+1	+1	+1	-1	-1
25	+1	+1	-1	-1	-1	+1	+1
26	+1	+1	-1	-1	+1	-1	+1
27	+1	+1	-1	+1	-1	+1	-1
28	+1	+1	-1	+1	+1	-1	-1
29	+1	+1	+1	-1	-1	-1	-1
30	+1	+1	+1	-1	+1	+1	-1

Number of Factors: 7.......................................**Design Name: FF0732**

(continued from previous page)

run	A	B	C	D	E	F	G
31	+1	+1	+1	+1	-1	-1	+1
32	+1	+1	+1	+1	+1	+1	+1
33	0	0	0	0	0	0	0
34	0	0	0	0	0	0	0
35	0	0	0	0	0	0	0

Confounding Pattern: Main effects (A through G) are not confounded with two-factor interactions. Two-factor interactions are not confounded with each other except for those listed below. Only one effect from each line below can be estimated. All two-factor interactions not listed below may be estimated.

D*E + F*G
D*F + E*G
E*F + D*G

* Include center points if possible (see Section 6).

Number of Factors: 8 ....Design Name: PB0812

Resolution:	III	Factorial:	2^8
Sample Size:	12	Fraction:	3/64

Factors

run	A	B	C	D	E	F	G	H
1	+1	-1	+1	-1	-1	-1	+1	+1
2	+1	+1	-1	+1	-1	-1	-1	+1
3	-1	+1	+1	-1	+1	-1	-1	-1
4	+1	-1	+1	+1	-1	+1	-1	-1
5	+1	+1	-1	+1	+1	-1	+1	-1
6	+1	+1	+1	-1	+1	+1	-1	+1
7	-1	+1	+1	+1	-1	+1	+1	-1
8	-1	-1	+1	+1	+1	-1	+1	+1
9	-1	-1	-1	+1	+1	+1	-1	+1
10	+1	-1	-1	-1	+1	+1	+1	-1
11	-1	+1	-1	-1	-1	+1	+1	+1
12	-1	-1	-1	-1	-1	-1	-1	-1

Confounding Pattern: All main effects (A through H) are partially confounded with two-factor interactions. Main effects are not confounded with other main effects.

Number of Factors: 8.........................Design Name: FF0816

Resolution:	IV	Factorial: 2^8
Sample Size:		Fraction: 1/16
Base Design:	16	
Center Points:	3*	

<u>Factors</u>

run	A	B	C	D	E	F	G	H
1	-1	-1	-1	-1	-1	-1	-1	-1
2	-1	-1	-1	+1	+1	+1	+1	-1
3	-1	-1	+1	-1	+1	+1	-1	+1
4	-1	-1	+1	+1	-1	-1	+1	+1
5	-1	+1	-1	-1	+1	-1	+1	+1
6	-1	+1	-1	+1	-1	+1	-1	+1
7	-1	+1	+1	-1	-1	+1	+1	-1
8	-1	+1	+1	+1	+1	-1	-1	-1
9	+1	-1	-1	-1	-1	+1	+1	+1
10	+1	-1	-1	+1	+1	-1	-1	+1
11	+1	-1	+1	-1	+1	-1	+1	-1
12	+1	-1	+1	+1	-1	+1	-1	-1
13	+1	+1	-1	-1	+1	+1	-1	-1
14	+1	+1	-1	+1	-1	-1	+1	-1
15	+1	+1	+1	-1	-1	-1	-1	+1
16	+1	+1	+1	+1	+1	+1	+1	+1
17	0	0	0	0	0	0	0	0
18	0	0	0	0	0	0	0	0
19	0	0	0	0	0	0	0	0
20	0	0	0	0	0	0	0	0

<u>Confounding Pattern</u>: Main effects (A through H) are not confounded with two factor interactions. Only one two-factor interaction from each line below can be estimated. (See Section 4.)

$$A*B + E*F + D*G + C*H$$
$$A*C + D*F + B*H + E*G$$
$$B*C + D*E + A*H + F*G$$
$$A*D + C*F + B*G + E*H$$
$$B*D + C*E + A*G + F*H$$
$$A*E + B*F + C*G + D*H$$
$$C*D + B*E + A*F + G*H$$

* Include center points if possible (see Section 6).

Number of Factors: 8 .... **Design Name: PB0824$^+$**

Resolution: IV	Factorial:	2^8
Sample Size:	Fraction:	3/32
Base Design: 24		
Center Points: 3*		

Factors

run	A	B	C	D	E	F	G	H
1	-1	+1	-1	+1	-1	-1	-1	+1
2	-1	+1	+1	-1	+1	-1	-1	-1
3	-1	-1	+1	+1	-1	+1	-1	-1
4	-1	+1	-1	+1	+1	-1	+1	-1
5	-1	+1	+1	-1	+1	+1	-1	+1
6	-1	+1	+1	+1	-1	+1	+1	-1
7	-1	-1	+1	+1	+1	-1	+1	+1
8	-1	-1	-1	+1	+1	+1	-1	+1
9	-1	-1	-1	-1	+1	+1	+1	-1
10	-1	+1	-1	-1	-1	+1	+1	+1
11	-1	-1	+1	-1	-1	-1	+1	+1
12	-1	-1	-1	-1	-1	-1	-1	-1
13	+1	-1	+1	-1	+1	+1	+1	-1
14	+1	-1	-1	+1	-1	+1	+1	+1
15	+1	+1	-1	-1	+1	-1	+1	+1
16	+1	-1	+1	-1	-1	+1	-1	+1
17	+1	-1	-1	+1	-1	-1	+1	-1
18	+1	-1	-1	-1	+1	-1	-1	+1
19	+1	+1	-1	-1	-1	+1	-1	-1
20	+1	+1	+1	-1	-1	-1	+1	-1
21	+1	+1	+1	+1	-1	-1	-1	+1
22	+1	-1	+1	+1	+1	-1	-1	-1
23	+1	+1	-1	+1	+1	+1	-1	-1
24	+1	+1	+1	+1	+1	+1	+1	+1
25	0	0	0	0	0	0	0	0
26	0	0	0	0	0	0	0	0
27	0	0	0	0	0	0	0	0

<u>Confounding Pattern</u>: Main effects are not confounded with two-factor interactions. All two-factor interactions are partially confounded with other two-factor interactions. All two-factor interactions with factor A can be estimated.

* Include center points if possible (see Section 6)

$^+$ This is a "fold-over" of a 12 run Plackett-Burman design. See the discussion in Section 5.

Number of Factors: 9 ....**Design Name: PB0912**

	Resolution:	III			Factorial:	2^9
	Sample Size:	12			Fraction:	3/128

Factors

run	A	B	C	D	E	F	G	H	I
1	+1	-1	+1	-1	-1	-1	+1	+1	+1
2	+1	+1	-1	+1	-1	-1	-1	+1	+1
3	-1	+1	+1	-1	+1	-1	-1	-1	+1
4	+1	-1	+1	+1	-1	+1	-1	-1	-1
5	+1	+1	-1	+1	+1	-1	+1	-1	-1
6	+1	+1	+1	-1	+1	+1	-1	+1	-1
7	-1	+1	+1	+1	-1	+1	+1	-1	+1
8	-1	-1	+1	+1	+1	-1	+1	+1	-1
9	-1	-1	-1	+1	+1	+1	-1	+1	+1
10	+1	-1	-1	-1	+1	+1	+1	-1	+1
11	-1	+1	-1	-1	-1	+1	+1	+1	-1
12	-1	-1	-1	-1	-1	-1	-1	-1	-1

Confounding Pattern: All main effects (A through I) are partially confounded with two-factor interactions. Main effects are not confounded with other main effects. No two-factor interactions should be estimated in this design.

Number of Factors: 9 ....Design Name: FF0916

	Resolution:	III			Factorial:	2^9		
	Sample Size:	16			Fraction:	1/32		

Factors

run	A	B	C	D	E	F	G	H	I
1	-1	-1	-1	-1	-1	-1	-1	-1	+1
2	-1	-1	-1	+1	+1	+1	+1	-1	-1
3	-1	-1	+1	-1	+1	+1	-1	+1	-1
4	-1	-1	+1	+1	-1	-1	+1	+1	+1
5	-1	+1	-1	-1	+1	-1	+1	+1	+1
6	-1	+1	-1	+1	-1	+1	-1	+1	-1
7	-1	+1	+1	-1	-1	+1	+1	-1	-1
8	-1	+1	+1	+1	+1	-1	-1	-1	+1
9	+1	-1	-1	-1	-1	+1	+1	+1	+1
10	+1	-1	-1	+1	+1	-1	-1	+1	-1
11	+1	-1	+1	-1	+1	-1	+1	-1	-1
12	+1	-1	+1	+1	-1	+1	-1	-1	+1
13	+1	+1	-1	-1	+1	+1	-1	-1	+1
14	+1	+1	-1	+1	-1	-1	+1	-1	-1
15	+1	+1	+1	-1	-1	-1	-1	+1	-1
16	+1	+1	+1	+1	+1	+1	+1	+1	+1

Confounding Pattern: Main effects are confounded with two-factor interactions. Only one effect from each line below can be estimated. The maximum number of effects that can be estimated is 15 (9 main effects and 6 interactions). (See Section 4.)

A + F*I
B + E*I
C + D*I
D + C*I
E + B*I
F + A*I
G + H*I
H + G*I
I + C*D + B*E + A*F + G*H
   A*B + E*F + D*G + C*H
   A*C + D*F + B*H + E*G
   B*C + D*E + A*H + F*G
   A*D + C*F + B*G + E*H
   B*D + C*E + A*G + F*H
   A*E + B*F + C*G + D*H

Number of Factors: 9.................................... **Design Name: PB0924[+]**

	Resolution:	IV		Factorial:	2^9
	Sample Size:			Fraction:	3/64
	Base Design:	24			
	Center Points:	3*			

Factors

run	A	B	C	D	E	F	G	H	I
1	-1	+1	-1	+1	-1	-1	-1	+1	+1
2	-1	+1	+1	-1	+1	-1	-1	-1	+1
3	-1	-1	+1	+1	-1	+1	-1	-1	-1
4	-1	+1	-1	+1	+1	-1	+1	-1	-1
5	-1	+1	+1	-1	+1	+1	-1	+1	-1
6	-1	+1	+1	+1	-1	+1	+1	-1	+1
7	-1	-1	+1	+1	+1	-1	+1	+1	-1
8	-1	-1	-1	+1	+1	+1	-1	+1	+1
9	-1	-1	-1	-1	+1	+1	+1	-1	+1
10	-1	+1	-1	-1	-1	+1	+1	+1	-1
11	-1	-1	+1	-1	-1	-1	+1	+1	+1
12	-1	-1	-1	-1	-1	-1	-1	-1	-1
13	+1	-1	+1	-1	+1	+1	+1	-1	-1
14	+1	-1	-1	+1	-1	+1	+1	+1	-1
15	+1	+1	-1	-1	+1	-1	+1	+1	+1
16	+1	-1	+1	-1	-1	+1	-1	+1	+1
17	+1	-1	-1	+1	-1	-1	+1	-1	+1
18	+1	-1	-1	-1	+1	-1	-1	+1	-1
19	+1	+1	-1	-1	-1	+1	-1	-1	+1
20	+1	+1	+1	-1	-1	-1	+1	-1	-1
21	+1	+1	+1	+1	-1	-1	-1	+1	-1
22	+1	-1	+1	+1	+1	-1	-1	-1	+1
23	+1	+1	-1	+1	+1	+1	-1	-1	-1
24	+1	+1	+1	+1	+1	+1	+1	+1	+1
25	0	0	0	0	0	0	0	0	0
26	0	0	0	0	0	0	0	0	0
27	0	0	0	0	0	0	0	0	0

Confounding Pattern: Main effects are not confounded with two-factor interactions. All two-factor interactions are partially confounded with other two-factor interactions. All two-factor interactions with factor A can be estimated.

* Include center points if possible (see Section 6).

[+] This is a "fold-over" of a 12 run Plackett-Burman design. See the discussion in Section 5.

Number of Factors: 10 ....Design Name: PB1012

	Resolution:	III		Factorial:	2^{10}
	Sample Size:	12		Fraction:	3/256

Factors

run	A	B	C	D	E	F	G	H	I	J
1	+1	-1	+1	-1	-1	-1	+1	+1	+1	-1
2	+1	+1	-1	+1	-1	-1	-1	+1	+1	+1
3	-1	+1	+1	-1	+1	-1	-1	-1	+1	+1
4	+1	-1	+1	+1	-1	+1	-1	-1	-1	+1
5	+1	+1	-1	+1	+1	-1	+1	-1	-1	-1
6	+1	+1	+1	-1	+1	+1	-1	+1	-1	-1
7	-1	+1	+1	+1	-1	+1	+1	-1	+1	-1
8	-1	-1	+1	+1	+1	-1	+1	+1	-1	+1
9	-1	-1	-1	+1	+1	+1	-1	+1	+1	-1
10	+1	-1	-1	-1	+1	+1	+1	-1	+1	+1
11	-1	+1	-1	-1	-1	+1	+1	+1	-1	+1
12	-1	-1	-1	-1	-1	-1	-1	-1	-1	-1

Confounding Pattern: All main effects (A through J) are partially confounded
with two-factor interactions. Main effects are not confounded with other
main effects. No two-factor interactions can be estimated in this design.

Number of Factors: 10 ....Design Name: FF1016

	Resolution:	III			Factorial:	2^{10}
	Sample Size:	16			Fraction:	1/64

Factors

run	A	B	C	D	E	F	G	H	I	J
1	-1	-1	-1	-1	-1	-1	-1	-1	+1	+1
2	-1	-1	-1	+1	+1	+1	+1	-1	-1	-1
3	-1	-1	+1	-1	+1	+1	-1	+1	-1	-1
4	-1	-1	+1	+1	-1	-1	+1	+1	+1	+1
5	-1	+1	-1	-1	+1	-1	+1	+1	+1	-1
6	-1	+1	-1	+1	-1	+1	-1	+1	-1	+1
7	-1	+1	+1	-1	-1	+1	+1	-1	-1	+1
8	-1	+1	+1	+1	+1	-1	-1	-1	+1	-1
9	+1	-1	-1	-1	-1	+1	+1	+1	+1	-1
710	+1	-1	-1	+1	+1	-1	-1	+1	-1	+1
11	+1	-1	+1	-1	+1	-1	+1	-1	-1	+1
12	+1	-1	+1	+1	-1	+1	-1	-1	+1	-1
13	+1	+1	-1	-1	+1	+1	-1	-1	+1	+1
14	+1	+1	-1	+1	-1	-1	+1	-1	-1	-1
15	+1	+1	+1	-1	-1	-1	-1	+1	-1	-1
16	+1	+1	+1	+1	+1	+1	+1	+1	+1	+1

Confounding Pattern: Main effects are confounded with two-factor interactions. Only one effect from each line below can be estimated. The maximum number of effects that can be estimated is 15 (10 main effects and 5 interactions). (See Section 4.)

A + F*I + E*J
B + E*I + F*J
C + D*I + G*J
D + C*I + H*J
E + B*I + A*J
F + A*I + B*J
G + C*J + H*I
H + D*J + G*I
I + C*D + B*E + A*F + G*H
J + A*E + B*F + C*G + D*H
   A*B + E*F + D*G + C*H + I*J
   A*C + D*F + B*H + E*G
   B*C + D*E + A*H + F*G
   A*D + C*F + B*G + E*H
   B*D + C*E + A*G + F*H

Number of Factors: 10 .... Design Name: PB1024$^+$

Resolution:	IV	Factorial:	2^{10}
Sample Size:		Fraction:	3/128
Base Design:	24		
Center Points:	3*		

Factors

run	A	B	C	D	E	F	G	H	I	J
1	-1	+1	-1	+1	-1	-1	-1	+1	+1	+1
2	-1	+1	+1	-1	+1	-1	-1	-1	+1	+1
3	-1	-1	+1	+1	-1	+1	-1	-1	-1	+1
4	-1	+1	-1	+1	+1	-1	+1	-1	-1	-1
5	-1	+1	+1	-1	+1	+1	-1	+1	-1	-1
6	-1	+1	+1	+1	-1	+1	+1	-1	+1	-1
7	-1	-1	+1	+1	+1	-1	+1	+1	-1	+1
8	-1	-1	-1	+1	+1	+1	-1	+1	+1	-1
9	-1	-1	-1	-1	+1	+1	+1	-1	+1	+1
10	-1	+1	-1	-1	-1	+1	+1	+1	-1	+1
11	-1	-1	+1	-1	-1	-1	+1	+1	+1	-1
12	-1	-1	-1	-1	-1	-1	-1	-1	-1	-1
13	+1	-1	+1	-1	+1	+1	+1	-1	-1	-1
14	+1	-1	-1	+1	-1	+1	+1	+1	-1	-1
15	+1	+1	-1	-1	+1	-1	+1	+1	+1	-1
16	+1	-1	+1	-1	-1	+1	-1	+1	+1	+1
17	+1	-1	-1	+1	-1	-1	+1	-1	+1	+1
18	+1	-1	-1	-1	+1	-1	-1	+1	-1	+1
19	+1	+1	-1	-1	-1	+1	-1	-1	+1	-1
20	+1	+1	+1	-1	-1	-1	+1	-1	-1	+1
21	+1	+1	+1	+1	-1	-1	-1	+1	-1	-1
22	+1	-1	+1	+1	+1	-1	-1	-1	+1	-1
23	+1	+1	-1	+1	+1	+1	-1	-1	-1	+1
24	+1	+1	+1	+1	+1	+1	+1	+1	+1	+1
25	0	0	0	0	0	0	0	0	0	0
26	0	0	0	0	0	0	0	0	0	0
27	0	0	0	0	0	0	0	0	0	0

Confounding Pattern: Main effects are not confounded with two-factor interactions. All two-factor interactions are partially confounded with other two-factor interactions. All two-factor interactions with factor A can be estimated.

* Include center points if possible (see Section 6).

$^+$ This is a "fold-over" of a 12 run Plackett-Burman design. See the discussion in Section 5.

Number of Factors: 11.....................................**Design Name: PB1112**

	Resolution:	III		Factorial:	2^{11}
	Sample Size:	12		Fraction:	3/512

Factors

run	A	B	C	D	E	F	G	H	I	J	K
1	+1	-1	+1	-1	-1	-1	+1	+1	+1	-1	+1
2	+1	+1	-1	+1	-1	-1	-1	+1	+1	+1	-1
3	-1	+1	+1	-1	+1	-1	-1	-1	+1	+1	+1
4	+1	-1	+1	+1	-1	+1	-1	-1	-1	+1	+1
5	+1	+1	-1	+1	+1	-1	+1	-1	-1	-1	+1
6	+1	+1	+1	-1	+1	+1	-1	+1	-1	-1	-1
7	-1	+1	+1	+1	-1	+1	+1	-1	+1	-1	-1
8	-1	-1	+1	+1	+1	-1	+1	+1	-1	+1	-1
9	-1	-1	-1	+1	+1	+1	-1	+1	+1	-1	+1
10	+1	-1	-1	-1	+1	+1	+1	-1	+1	+1	-1
11	-1	+1	-1	-1	-1	+1	+1	+1	-1	+1	+1
12	-1	-1	-1	-1	-1	-1	-1	-1	-1	-1	-1

Confounding Pattern: All main effects (A through K) are partially confounded with two-factor interactions. Main effects are not confounded with other main effects. (See Section 4.) No two-factor interactions can be estimated in this design.

Number of Factors: 11 ....**Design Name: FF1116**

| | | | | Resolution: | III | | | Factorial: | | 2^{11} |
| | | | | Sample Size: | 16 | | | Fraction: | | 1/128 |

Factors

run	A	B	C	D	E	F	G	H	I	J	K
1	-1	-1	-1	-1	-1	-1	-1	-1	+1	+1	+1
2	-1	-1	-1	+1	+1	+1	+1	-1	-1	-1	-1
3	-1	-1	+1	-1	+1	+1	-1	+1	-1	-1	+1
4	-1	-1	+1	+1	-1	-1	+1	+1	+1	+1	-1
5	-1	+1	-1	-1	+1	-1	+1	+1	+1	-1	-1
6	-1	+1	-1	+1	-1	+1	-1	+1	-1	+1	+1
7	-1	+1	+1	-1	-1	+1	+1	-1	-1	+1	-1
8	-1	+1	+1	+1	+1	-1	-1	-1	+1	-1	+1
9	+1	-1	-1	-1	-1	+1	+1	+1	+1	-1	+1
10	+1	-1	-1	+1	+1	-1	-1	+1	-1	+1	-1
11	+1	-1	+1	-1	+1	-1	+1	-1	-1	+1	+1
12	+1	-1	+1	+1	-1	+1	-1	-1	+1	-1	-1
13	+1	+1	-1	-1	+1	+1	-1	-1	+1	+1	-1
14	+1	+1	-1	+1	-1	-1	+1	-1	-1	-1	+1
15	+1	+1	+1	-1	-1	-1	-1	+1	-1	-1	-1
16	+1	+1	+1	+1	+1	+1	+1	+1	+1	+1	+1

Confounding Pattern: Main effects are confounded with two-factor interactions. Only one effect from each line below can be estimated. The maximum number of effects that can be estimated is 15 (11 main effects and 4 interactions). (See Section 4.)

A + F*I + E*J + G*K
B + E*I + D*K + F*J
C + D*I + E*K + G*J
D + C*I + B*K + H*J
E + B*I + A*J + C*K
F + A*I + B*J + H*K
G + A*K + C*J + H*I
H + D*J + G*I + F*K
I + C*D + B*E + A*F + G*H
J + A*E + B*F + C*G + D*H
K + B*D + C*E + A*G + F*H
   A*B + E*F + D*G + C*H + I*J
   A*C + D*F + B*H + E*G + J*K
   B*C + D*E + A*H + F*G + I*K
   A*D + C*F + B*G + E*H

Number of Factors: 11 .... **Design Name: PB1124[+]**

	Resolution:	IV		Factorial:	2^{11}
	Sample Size:			Fraction:	3/256
	Base Design:	24			
	Center Points:	3*			

Factors

run	A	B	C	D	E	F	G	H	I	J	K
1	-1	+1	-1	+1	-1	-1	-1	+1	+1	+1	-1
2	-1	+1	+1	-1	+1	-1	-1	-1	+1	+1	+1
3	-1	-1	+1	+1	-1	+1	-1	-1	-1	+1	+1
4	-1	+1	-1	+1	+1	-1	+1	-1	-1	-1	+1
5	-1	+1	+1	-1	+1	+1	-1	+1	-1	-1	-1
6	-1	+1	+1	+1	-1	+1	+1	-1	+1	-1	-1
7	-1	-1	+1	+1	+1	-1	+1	+1	-1	+1	-1
8	-1	-1	-1	+1	+1	+1	-1	+1	+1	-1	+1
9	-1	-1	-1	-1	+1	+1	+1	-1	+1	+1	-1
10	-1	+1	-1	-1	-1	+1	+1	+1	-1	+1	+1
11	-1	-1	+1	-1	-1	-1	+1	+1	+1	-1	+1
12	-1	-1	-1	-1	-1	-1	-1	-1	-1	-1	-1
13	+1	-1	+1	-1	+1	+1	+1	-1	-1	-1	+1
14	+1	-1	-1	+1	-1	+1	+1	+1	-1	-1	-1
15	+1	+1	-1	-1	+1	-1	+1	+1	+1	-1	-1
16	+1	-1	+1	-1	-1	+1	-1	+1	+1	+1	-1
17	+1	-1	-1	+1	-1	-1	+1	-1	+1	+1	+1
18	+1	-1	-1	-1	+1	-1	-1	+1	-1	+1	+1
19	+1	+1	-1	-1	-1	+1	-1	-1	+1	-1	+1
20	+1	+1	+1	-1	-1	-1	+1	-1	-1	+1	-1
21	+1	+1	+1	+1	-1	-1	-1	+1	-1	-1	+1
22	+1	-1	+1	+1	+1	-1	-1	-1	+1	-1	-1
23	+1	+1	-1	+1	+1	+1	-1	-1	-1	+1	-1
24	+1	+1	+1	+1	+1	+1	+1	+1	+1	+1	+1
25	0	0	0	0	0	0	0	0	0	0	0
26	0	0	0	0	0	0	0	0	0	0	0
27	0	0	0	0	0	0	0	0	0	0	0

Confounding Pattern: Main effects are not confounded with two-factor interactions. All two-factor interactions are partially confounded with other two-factor interactions. All two-factor interactions with factor A can be estimated.

* Include center points if possible (see Section 6).

[+] This is a "fold-over" of a 12 run Plackett-Burman design. See the discussion in Section 5.

4.   Estimating effects in a saturated design

A design is effect saturated if the total number of estimated effects is equal to n-1 (where n is the number of runs). Some designs in the digest are effect saturated when the maximum number of effects are included in the model. Each effect estimated uses 1 degree of freedom and 1 degree of freedom is used to estimate the constant term, so when a design is effect saturated, there are no degrees of freedom remaining to estimate error.

Some PC software packages will not allow the estimation of all effects simultaneously for an effect saturated design. In this case, estimate the effects after excluding 1 effect from the model initially. Then drop one of the least significant effects from the model and include the effect that was originally excluded to get estimates of all of the effects. In an orthogonal design, the estimates of the effects will not change as other terms are changed in the model. This is not true for a nonorthogonal design.

Standard tests of hypotheses can not be used with an unreplicated, effect saturated design because there is no independent estimate of the experimental error. Consequently, we recommend the use of the graphical methods described in Chapter 4 for the analysis of unreplicated, effect saturated designs.

5.   Resolution of irregular and mixed-level fractional factorial designs

The standard fractional factorial designs of 8, 16 and 32 runs included in this digest are used extensively and their properties are well understood. Unfortunately, there are times when the properties of the 8 run designs are unsatisfactory, but a 16 run design is not feasible. In order to fill this gap and provide a set of designs which is as complete and as useful as possible, a number of 12 run designs are included. Similarly, 24 run designs are included which tend to fill the gap between the 16 run and 32 run designs. Some of the 12 run and 24 run designs have one factor at three levels.

In fractional factorial designs, effects are either orthogonal to each other (unconfounded) or they are completely confounded. In this framework, the concept of resolution is well defined (see the definitions in Section 1). Most of the 12 run and 24 run designs contain some partial confounding. Because the partial confounding for these designs is quite limited, the patterns are not shown in the Design Digest. These designs are called *nearly* Resolution IV or V as appropriate.

Plackett-Burman designs

Five 12 run Plackett-Burman designs have been included. These designs are useful when there are many factors, a 16 run design is not feasible, and main effects are of primary interest. The Plackett-Burman designs included have from 7 to 11 factors and are all of Resolution III. These particular designs were given in Box, Hunter and Hunter (1978), but are originally due to Plackett and Burman (1946).

Main effects in these designs are unconfounded with each other. Main effects are partially confounded (correlation = ±.33) with all two factor interactions which do not contain the main effect factor. For example, A is partially confounded with B*C, but A is unconfounded with A*B. Similarly, pairs of two-factor interactions are partially confounded (correlation = .33) when they do not share a letter. For example, A*B is partially confounded with C*D, but A*B and A*C are unconfounded.

In general, no two-factor interactions should be estimated in a Plackett-Burman design. However, the additional factors from the 11 factor Plackett-Burman design can be included as dummy variables. If one of these dummy variables has an important effect on the response, then a large two-factor interaction is suspected. (See Chapter 9.)

Irregular fractional factorial designs

Three 12 run designs, IF0412, IF0512, and IF0612, due to John (1969) have been included in the digest. If a 16 run design is not feasible, these designs achieve a higher resolution than the 8 run fractional factorial designs. The 4 factor design is a three-quarter fraction of a 2^4 factorial and is nearly Resolution V. The 5 factor and 6 factor designs are nearly Resolution IV.

These designs can be analyzed in a straight-forward way using a statistical computer package. An alternative analysis method is given in Diamond (1981) and John (1969). John's method involves analyzing three subsets of the 12 runs in order to find all of the estimates of effects. When this method of analysis is used, much of the partial confounding is avoided. On the other hand, if one accepts the partial confounding in the entire 12 run design and analyzes the experiment in the usual way, the identical estimates can be achieved. If the maximum number of interactions are included, the two sets of estimates will match exactly. For this reason, the experimenter should typically be satisfied with the straight-forward analysis using a statistical computer package.

IF0624 is an irregular 3/8 fraction of a 2^6 factorial which was developed by Roger Liddle. This design has some partial confounding, but it allows the estimation of all main effects and two-factor interactions. It is nearly Resolution V. (See the discussion of this design in Section 4 of Chapter 7 for more information about the partial confounding structure.)

Mixed-level fractional factorial designs

Designs MF0412, MF0512, and MF0524 are mixed-level fractional factorial designs (Liddle and Haaland, 1988). Each of these designs allows for one three-level factor.

MF0412 is a 1/2 fraction of a $3*2^3$ factorial. MF0512 is a 1/4 fraction of a $3*2^4$ factorial. MF0524 is a 1/2 fraction of a $3*2^4$ factorial. MF0412 and MF0524 are nearly Resolution V whereas MF0512 is nearly Resolution IV. Only MF0524 allows interactions between the quadratic effect and other main effects (for example, A^2*B, etc.).

At least initially, a quadratic effect for the three level factor should be included in the model. (See Section 7 of the Design Digest and Section 1 of Chapter 7 for how to correctly define the quadratic effect.) When the three level factor is qualitative, the quadratic effect does not have its usual interpretation. However, it is still needed to distinguish among different responses at the 3 levels of the qualitative factor. In this case, an interaction plot provides the most reliable interpretation of the differences among levels of the qualitative factor (see Chapter 7).

Fold-over Plackett-Burman designs

Five 24 run "fold-over" designs derived from Plackett-Burman designs are included in the digest. A 12 run Plackett-Burman design of Resolution III can be folded-over by repeating the 12 runs with all ones and minus ones reversed (see Box and Wilson, 1951). This results in a 24 run design of Resolution IV. The Design Digest includes designs of this type for 7 to 11 factors.

Main effects in these designs are unconfounded with other main effects or two-factor interactions. Two-factor interactions are partially confounded (correlation = $\pm.33$) with any two factor interactions which do not have a letter in common. For example, A*B is unconfounded with A*C, but A*B and C*D are partially confounded. Thus, the only two-factor interactions which can be estimated are those between factor A and the other main effects. For example, using PB1024 there is no partial confounding among the following effects: A, B, C, ..., J, A*B, A*C, ..., A*J.

6.   Center points in screening designs

Center points are recommended for most of the designs in the digest, (8 run designs with 3-4 factors, 16 run designs with 4-8 factors, 24 run designs with 4-11 factors, and 32 run designs with 5-7 factors). Center points provide information about the interior of the experimental region, and they allow you to check for curvature. Center points also provide additional degrees of freedom for error which result in greater power when testing the significance of effects. For these reasons, the recommended center points should typically be included in the experiment.

Center points have not been included in designs of Resolution III. Typically designs of Resolution III are used when there are severe sample size restrictions and information about a large number of factors is desired. Under these circumstances, center points are of less value and are difficult or impossible to include. If there is not a severe sample size restriction, consider running a larger experiment in order to achieve Resolution IV or V.

Center points have not been included in 12 run designs for similar reasons. If there are not strong reasons for limiting the number of runs to 12, consider using a 16 run experiment.

Under certain circumstances, it is reasonable to delete recommended center points from a design. If the additional runs would make the experiment very costly or impossible to complete, then removing some or all of the center points may be necessary. Including center points will never hurt the statistical properties of a properly analyzed experiment, but the center points are of reduced value under the following conditions. First, when information about the center of the experimental region is not needed and the experimenter is not interested in possible curvature. Second, when the design already has sufficient degrees of freedom for error.

Center points cannot be used if one or more of the factors are qualitative. In such a situation, center points are typically deleted from the design. If there are an even number of center points and a single qualitative factor, it is possible to replace the center points with face centered points. Initially the center points will have all factor levels set to zero. The zeros for the qualitative factor can be replaced by ones and minus ones, (half ones and half minus ones). The resulting design has no true center points, but the extra degrees of freedom for error are still available and a curvature check for the quantitative factors is still possible.

Another possibility for "adjusting" the center points would be to replace all the zeros (for the qualitative factor) by a single number (either a one or a minus one). This has the advantage of working for any number of center points, and could be used for more than one qualitative factor. On the other hand, it will affect certain orthogonality and optimality properties of the design. Such effects can be minimized by including a quadratic effect for one of the quantitative factors (as a curvature check) and leaving it in the model regardless of the significance level. This will restore the orthogonality of all remaining estimates.

Either of these techniques for "adjusting" the center points, is straight forward to implement. However, it is important to note that both techniques are *ad hoc* procedures.

7. Tests for curvature

When center points are included in a screening experiment, curvature in the response variables can be detected. If one or more quadratic effects are non-zero, then the center points may have higher (or lower) values of the response variables than would be expected from the remainder of the data points.

If curvature is significant, and no adjustment is made in the analysis of the experiment, then the estimate of error based on linear and interaction terms only will be artificially inflated. Drawing meaningful conclusions from the experiment will become more difficult. When non-zero quadratic effects are suspected and a larger sample size is feasible, a response surface experiment should be considered. Screening experiments are not designed to estimate and test individual quadratic effects, but there are ways of testing for curvature when center points are run.

A procedure to test for curvature in a screening experiment can be adapted from Myers (1976). Let $\bar{y}_1$ be the average of the n_1 response values from the base design. Let $\bar{y}_2$ be the average of the response values from the n_2 center points. A\t-test with n_2-1 degrees of freedom can then be used to compare these two averages. While this procedure is relatively simple and straightforward, it does involve an added level of complexity for the experimenter.

A simple alternative procedure for testing curvature is presented in Chapter 7. This method is generally equivalent to Myers' method and can be performed during a typical statistical analysis of the data. Include a quadratic term in the model, ("1.5*factorA*factorA-1", where "factorA" is the name of one of the factors). If the p-value for the resulting estimate is small, say, less than .2 perhaps, then one or more factors probably have a non-zero quadratic effect. In this situation, the quadratic term should be left in the model so that the error sum of squares is not artificially inflated. The estimate for this quadratic term should NOT be interpreted as a quadratic effect for a particular factor, since all the quadratic effects are confounded with each other.

8.   Quick reference tables for response surface experiments

Sample sizes for given numbers of factors

Table 5 lists the recommended sample sizes for central composite or face centered cube designs for a given number of factors.

Table 5. Recommended Sample Sizes for Response Surface Designs

Number Factors	Sample Size
2	11
3	18
4	28
5	30

* These sample sizes include the recommended center points. See Section 10 for further information on center points in Response Surface experiments.

Design names

Table 6 can be used to identify the appropriate design name.

Table 6. Names of Response Surface Designs

Number of Factors	Central Composite	Face Centered Cube
2	CC0211	FC0211
3	CC0318	FC0318
4	CC0428	FC0428
5	CC0530	FC0530

Note: The design name for the 3 factor Box-Behnken design is BB0316.

9. Listings of designs for response surface experiments

This section includes central composite designs, face centered cube designs, and a Box-Behnken design. All the designs in this section allow the estimation of all main effects, all two-factor interactions, and quadratic effects for each factor.

The designs are presented in the following order:

- Number of factors: from 2 to 5.
- Design: for each number of factors, the central composite design is presented first, followed by the face centered cube design.

Box-Behnken Designs are available for 3 or more factors, but we have included only the 3 factor Box-Behnken design. For a discussion of the use of center points in response surface experiments, see Section 10 of this digest.

Number of Factors: 2......................................Design Name: CC0211

Resolution:	V	Type: Central Composite
Sample Size:	11	

	Factors	
run	A	B
1	-1	-1
2	-1	+1
3	+1	-1
4	+1	+1
5	-1.41	0
6	+1.41	0
7	0	-1.41
8	0	+1.41
9	0	0
10	0	0
11	0	0

Confounding Pattern: All main effects, quadratic effects, and two-factor inter-actions can be estimated.

Center points must be included in this design (see Section 10).

Number of Factors: 2 .... Design Name: FC0211

Resolution:	V	Type:	Face Centered Cube
Sample Size:	11		

Factors

run	A	B
1	-1	-1
2	-1	+1
3	+1	-1
4	+1	+1
5	-1	0
6	+1	0
7	0	-1
8	0	+1
9	0	0
10	0	0
11	0	0

Confounding Pattern: All main effects, quadratic effects, and two-factor interactions can be estimated.

Center points must be included in this design (see Section 10).

Number of Factors: 3.................................... **Design Name: BB0316**

	Resolution:	V	Type:	Box-Behnken
	Sample Size:	16		

Factors

run	A	B	C
1	-1	-1	0
2	-1	+1	0
3	+1	-1	0
4	+1	+1	0
5	-1	0	-1
6	-1	0	+1
7	+1	0	-1
8	+1	0	+1
9	0	-1	-1
10	0	-1	+1
11	0	+1	-1
12	0	+1	+1
13	0	0	0
14	0	0	0
15	0	0	0
16	0	0	0

Confounding Pattern: All main effects, quadratic effects, and two factor interactions can be estimated.

Center points must be included in this design (see Section 10).

Number of Factors: 3.......................................**Design Name: CC0318**

Resolution:	V	Type:	Central Composite
Sample Size:	18		

Factors

run	A	B	C
1	-1	-1	-1
2	-1	-1	+1
3	-1	+1	-1
4	-1	+1	+1
5	+1	-1	-1
6	+1	-1	+1
7	+1	+1	-1
8	+1	+1	+1
9	-1.68	0	0
10	+1.68	0	0
11	0	-1.68	0
12	0	+1.68	0
13	0	0	-1.68
14	0	0	+1.68
15	0	0	0
16	0	0	0
17	0	0	0
18	0	0	0

Confounding Pattern: All main effects, quadratic effects, and two factor interactions can be estimated.

Center points must be included in this design (see Section 10).

Number of Factors: 3..................................... **Design Name: FC0318**

	Resolution:	V	Type:	Face Centered Cube
	Sample Size:	18		

Factors

run	A	B	C
1	-1	-1	-1
2	-1	-1	+1
3	-1	+1	-1
4	-1	+1	+1
5	+1	-1	-1
6	+1	-1	+1
7	+1	+1	-1
8	+1	+1	+1
9	-1	0	0
10	+1	0	0
11	0	-1	0
12	0	+1	0
13	0	0	-1
14	0	0	+1
15	0	0	0
16	0	0	0
17	0	0	0
18	0	0	0

Confounding Pattern: All main effects, quadratic effects, and two factor interactions can be estimated.

Center points must be included in this design (see Section 10).

Number of Factors: 4.....................................**Design Name: CC0428**

Resolution:	V			Type:	Central Composite		
Sample Size:	28						

	Factors					Factors		
run	A	B	C	D	run	A	B	C	D
1	-1	-1	-1	-1	17	-2	0	0	0
2	-1	-1	-1	+1	18	+2	0	0	0
3	-1	-1	+1	-1	19	0	-2	0	0
4	-1	-1	+1	+1	20	0	+2	0	0
5	-1	+1	-1	-1	21	0	0	-2	0
6	-1	+1	-1	+1	22	0	0	+2	0
7	-1	+1	+1	-1	23	0	0	0	-2
8	-1	+1	+1	+1	24	0	0	0	+2
9	+1	-1	-1	-1	25	0	0	0	0
10	+1	-1	-1	+1	26	0	0	0	0
11	+1	-1	+1	-1	27	0	0	0	0
12	+1	-1	+1	+1	28	0	0	0	0
13	+1	+1	-1	-1					
14	+1	+1	-1	+1					
15	+1	+1	+1	-1					
16	+1	+1	+1	+1					

Confounding Pattern: All main effects, quadratic effects, and two factor interactions can be estimated.

Center points must be included in this design (see Section 10).

Number of Factors: 4..................................... **Design Name: FC0428**

	Resolution:	V		Type:	Face Centered Cube
	Sample Size:	28			

		Factors					Factors		
run	A	B	C	D	run	A	B	C	D
1	-1	-1	-1	-1	17	-1	0	0	0
2	-1	-1	-1	+1	18	+1	0	0	0
3	-1	-1	+1	-1	19	0	-1	0	0
4	-1	-1	+1	+1	20	0	+1	0	0
5	-1	+1	-1	-1	21	0	0	-1	0
6	-1	+1	-1	+1	22	0	0	+1	0
7	-1	+1	+1	-1	23	0	0	0	-1
8	-1	+1	+1	+1	24	0	0	0	+1
9	+1	-1	-1	-1	25	0	0	0	0
10	+1	-1	-1	+1	26	0	0	0	0
11	+1	-1	+1	-1	27	0	0	0	0
12	+1	-1	+1	+1	28	0	0	0	0
13	+1	+1	-1	-1					
14	+1	+1	-1	+1					
15	+1	+1	+1	-1					
16	+1	+1	+1	+1					

Confounding Pattern: All main effects, quadratic effects, and two factor interactions can be estimated.

Center points must be included in this design (see Section 10).

Number of Factors: 5..**Design Name: CC0530**

	Resolution:	V		Type:	Central Composite
	Sample Size:	30			

		Factors							Factors		
run	A	B	C	D	E	run	A	B	C	D	E
1	-1	-1	-1	-1	+1	17	-2	0	0	0	0
2	-1	-1	-1	+1	-1	18	+2	0	0	0	0
3	-1	-1	+1	-1	-1	19	0	-2	0	0	0
4	-1	-1	+1	+1	+1	20	0	+2	0	0	0
5	-1	+1	-1	-1	-1	21	0	0	-2	0	0
6	-1	+1	-1	+1	+1	22	0	0	+2	0	0
7	-1	+1	+1	-1	+1	23	0	0	0	-2	0
8	-1	+1	+1	+1	-1	24	0	0	0	+2	0
9	+1	-1	-1	-1	-1	25	0	0	0	0	-2
10	+1	-1	-1	+1	+1	26	0	0	0	0	+2
11	+1	-1	+1	-1	+1	27	0	0	0	0	0
12	+1	-1	+1	+1	-1	28	0	0	0	0	0
13	+1	+1	-1	-1	+1	29	0	0	0	0	0
14	+1	+1	-1	+1	-1	30	0	0	0	0	0
15	+1	+1	+1	-1	-1						
16	+1	+1	+1	+1	+1						

Confounding Pattern:All main effects, quadratic effects, and two factor interactions can be estimated.

Center points must be included in this design (see Section 10).

Number of Factors: 5................................... **Design Name: FC0530**

	Resolution:	V		Type:		Face Centered Cube
	Sample Size:	30				

			Factors						Factors		
run	A	B	C	D	E	run	A	B	C	D	E
1	-1	-1	-1	-1	+1	17	-1	0	0	0	0
2	-1	-1	-1	+1	-1	18	+1	0	0	0	0
3	-1	-1	+1	-1	-1	19	0	-1	0	0	0
4	-1	-1	+1	+1	+1	20	0	+1	0	0	0
5	-1	+1	-1	-1	-1	21	0	0	-1	0	0
6	-1	+1	-1	+1	+1	22	0	0	+1	0	0
7	-1	+1	+1	-1	+1	23	0	0	0	-1	0
8	-1	+1	+1	+1	-1	24	0	0	0	+1	0
9	+1	-1	-1	-1	-1	25	0	0	0	0	-1
10	+1	-1	-1	+1	+1	26	0	0	0	0	+1
11	+1	-1	+1	-1	+1	27	0	0	0	0	0
12	+1	-1	+1	+1	-1	28	0	0	0	0	0
13	+1	+1	-1	-1	+1	29	0	0	0	0	0
14	+1	+1	-1	+1	-1	30	0	0	0	0	0
15	+1	+1	+1	-1	-1						
16	+1	+1	+1	+1	+1						

Confounding Pattern: All main effects, quadratic effects, and two factor interactions can be estimated.

Center points must be included in this design (see Section 10).

10. Center points in response surface designs

Center points are included in every response surface design in this digest. There is some flexibility in the number of center points to run, but at least one center point must be run and multiple center points are strongly recommended.

Additional center points improve estimates of quadratic effects, and they provide additional degrees of freedom for error. They also provide extra information about the center of the experimental region where the best response values are often located.

For most applications of response surface experiments, all the center points which appear in the design should be used. If necessary, a single center point can be deleted without seriously harming the properties of the design.

References and Bibliography

Box, G. E. P. and N. R. Draper (1987). *Empirical Model-Building and Response Surfaces*. New York:Wiley

Box, G. E. P. and K. B. Wilson (1951). On the Experimental Attainment of Optimum Conditions. *J. Royal Statistical Society, Series B*, **13**, 1-X.

Box, G. E. P., W. G. Hunter, and J. S. Hunter (1978). *Statistics for Experimenters*. New York:Wiley.

Diamond, W. J. (1981). *Practical Experimental Designs*. Belmont, CA: Lifetime Learning Publications.

John, P. W. M. (1969). Some Non-orthogonal Fractions of 2^n Designs. *J. Royal Statistical Society, Series B*, **31**, 270-275.

Khuri, A. I., and J. A. Cornell (1987). *Response Surfaces: Designs & Analyses*. New York

Liddle, R. F. and P. D. Haaland (1988). Efficient nonstandard screening designs. Presented at the Annual Meetings of the American Statistical Association, New Orleans, LA.

Myers, R. H. (1976). *Response Surface Methodology*. Boston: Allyn and Bacon. (Reprinted by Edwards Bros., Ann Arbor, MI).

Plackett, R. L. and J. P. Burman (1946). The design of optimum multifactorial experiments. *Biometrika*, **33**, 305-325.

Wheeler, D. J. (1988). *Understanding Industrial Experimentation*. Knoxville, Tennessee: Statistical Process Control, Inc.

Index

Printed in the United States
122773LV00003B/121/A